共建万物和谐的美丽家园

To Build a Harmonious and Beautiful Homeland for All Creatures

滇金丝猴 Black Snub-nosed Monkey　奚志农 野性中国 / 摄

中国的生物多样性保护

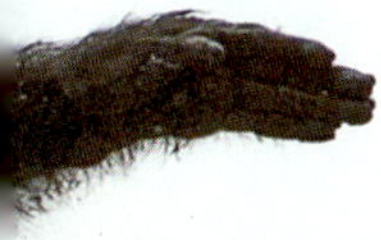

BIODIVERSITY CONSERVATION IN CHINA

中华人民共和国生态环境部　编著

Ministry of Ecology and Environment of
the People's Republic of China

中国环境出版集团・北京

大熊猫被誉为“中国国宝”，其憨态可掬的形象深受全世界人民的喜爱。在中国政府的大力保护下，大熊猫野生种群及繁育种群数量不断上升。

Giant panda, honored as the nation's treasure, has won world-wide popularity with its extraordinary adorableness. The wild population and captive breeding population of this species see a steady growth, thanks to the unreserved conservation efforts that Chinese government devoted.

冯江 自然影像中国 / 摄

在中国政府、科研院校及民间机构的大力保护下，朱鹮从野外最后的 7 只，发展到如今拥有壮大的野生种群和人工繁育种群，是野生动物保护史上最成功的案例之一。

With joint efforts of Chinese government, academic institutions and civil societies, both wild population and artificially-bred population of crested ibis have made a significant surge from merely 7, making crested ibis one of the most remarkable successes in the history of wildlife conservation.

赵纳勋 / 摄

中国政府高度重视藏羚的保护工作，在中国政府的呼吁下，国际社会通力合作保护藏羚，随着“沙图什贸易”的终止以及相应保护工作的不断完善，藏羚的种群数量得到有效恢复。

The Chinese government pays high attention to the conservation of Chirus. With its appeals, globally joint efforts have been devoted to save the species. As ‘Shahtoosh trading’ ceased, and relevant conservative work continuously improved, the population of Chirus has been remarkably restored.

吴秀山 / 摄

中国分布有 9 种鹤，丹顶鹤是最知名的一种，具有重要的文化意义。中国是丹顶鹤重要的繁殖地和越冬地，已在其主要分布区建立多个国家级自然保护区。

Among the 9 crane species distributing across China, red-crowned crane receives the most recognition and is regarded as a significant symbol of culture. With important breeding and wintering grounds of the crane located in the country, China has established several national nature reserves that cover the habitats of this species.

吴秀山 / 摄

高黎贡白眉长臂猿曾被认为是东白眉长臂猿，于 2017 年宣布为独立种。其主要种群栖息于高黎贡山等自然保护区内。

Skywalker hoolock gibbon, once misidentified as East hoolock gibbon, was declared as independent species in 2017. The majority of its population take shelter in several national reserves, such as Gaoligong Mountain.

赵超 / 摄

序

共建万物和谐的美丽世界

人类的生存与发展，每一步都离不开大自然的恩赐。生物多样性是人类赖以生存和发展的基础，是地球生命共同体的血脉和根基。然而，当今世界正面临生物多样性丧失和环境破坏的生态危机。人类活动改变了 75% 的陆地表面，影响了 66% 的海洋环境，超过 85% 的湿地已经丧失，25% 的物种正遭受灭绝威胁，近 1/5 的地球表面面临动植物入侵风险。

面对生态问题的挑战，人类是一荣俱荣、一损俱损的命运共同体。中国是世界上生物多样性最丰富的国家之一，2021 年 10 月 11–15 日，以“生态文明：共建地球生命共同体”为主题在中国昆明召开的《生物多样性公约》第十五次缔约方大会（COP15）期间，国家主席习近平在领导人峰会上强调，人不负青山，青山定不负人。人与自然应和谐共生，国际社会要携手同行，开启人类高质量发展新征程。要以生态文明建设为引领，协调人与自然关系；以绿色转型为驱动，助力全球可持续发展；以人民福祉为中心，促进社会公平正义；以国际法为基础，维护公平合理的国际治理体系。COP15 大会通过了《昆明宣言》，为全球环境治理注入新动力。中国将率先出资 15 亿元人民币成立昆明生物多样性基金，正式设立第一批国家公园，构建碳达峰、碳中和“1+N”政策体系。

中国始终坚持用生态文明理念指导发展，高度重视生物多样性保护，采取了有力行动，工作成效显著。

生物多样性战略地位明显提升。将“生态文明”写入宪法，成立中国生物多样性保护国家委员会，实施《中国生物多样性保护战略与行动计划（2011—2030 年）》，开展“联合国生物多样性十年中国行动”系列活动，实施生物多样性保护重大工程，加强生态保护和修复。

政策与法律法规体系加快建立。先后出台《生态文明体制改革总体方案》《关于进一步加强生物多样性保护的意见》《关于划定并严守生态保护红线的若干意见》《关于健全生态保护补偿机制的意见》等多项生物多样性保护相关改革举措，将生物多样性纳入经济社会发展、生态保护修复和国土空间相关规划。制定《生物安全法》，发布全面禁止非法野生动物交易的禁令。2021 年起，长江流域重点水域开始实行为期 10 年的禁捕政策。

生态空间保护力度持续加大。建立生态保护红线制度，初步划定的生态保护红线面积不小于陆域国土面积的 25%，覆盖重点生态功能区、生态环境敏感区和脆弱区等重要生态区域，有效保护了生物物种及其栖息环境。建设以国家公园为主体的自然保护地体系，各级各类自然保护地约 1.18 万个，占陆域国土面积的 18%、海域面积的 4.1%。

生物多样性保护成果丰硕。过去 10 年，中国森林资源增长面积超过 7 000 万公顷，居全球首位。长时间、大规模治理沙化、荒漠化，有效保护修复湿地，生物遗传资源收集保藏量位居世界前列。90% 的陆地生态系统类型、85% 的重点野生动物种群和 65% 的高等植物群落得到有效保护。珍稀濒危物种实现恢复性增长，朱鹮野外种群和人工繁育种群总数超过 6 600 只，大熊猫野生种群超过 1 800 只，受威胁程度等级由濒危降为易危，近百种极小种群野生植物得到抢救性保护。

国际交流与合作取得积极进展。坚定支持和践行多边主义，认真履行生态环境相关国际公约义务，全球生物多样性“爱知目标”的中国实践取得显著进展，20 个“爱知目标”完成情况均优于国际总体水平。宣布了二氧化碳排放力争于 2030 年前达到峰值、努力争取 2060 年前实现碳中和的承诺。倡议成立“一带一路”绿色发展国际联盟，与百余个国家开展环保交流合作，稳步推进《生物多样性公约》履约工作，与国际社会共商 2020 年后全球生物多样性保护战略。

展望未来，中国将从保护自然中寻找发展机遇，加快构筑保护优先、绿色发展的生态经济体系。完善生物多样性调查观测技术标准体系和保护成效评估体系。持续实施生物多样性保护重大工程，提升生态系统质量和稳定性。全面加强生物多样性保护与监管能力，严控重要生态空间用途改变，加强对野生动植物资源利用的全过程监管。提高社会各界保护生物多样性的自觉性和参与度。深度参与《生物多样性公约》进程，提升应对全球环境挑战行动力。

中国愿与国际社会携手同行，站在对人类文明负责任的高度，尊重自然、顺应自然、保护自然，落实《2030 年可持续发展议程》，为了地球生命共同体，探索人与自然和谐共生之路，共建清洁健康、生机勃勃、繁荣美丽的世界。

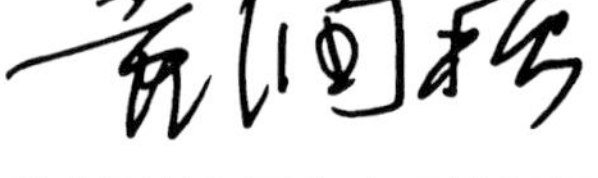

中华人民共和国生态环境部部长

Preface

To Build a Harmonious and Beautiful World for All Creatures

Every step of human survival and development is inseparable from the gift of nature. Biodiversity is the basis for human survival and development, and is the bloodline and foundation of the community of all life on Earth. However, the world today is facing the ecological crises of biodiversity loss and environmental destruction. Human activities have altered 75% of the land surface and affected 66% of the marine environment. More than 85% of wetlands have been lost, 25% of species are threatened with extinction, and nearly one-fifth of the earth's surface is at risk of invasive species.

Faced with the challenge of ecological problems, humanity is a community with a shared future. China is one of the countries with mega biodiversity. On October 11-15, 2021, the 15th Meeting of the Conference of the Parties to the Convention on Biological Diversity (COP15) was held in Kunming, China, under the theme of "Ecological Civilization: Building a Shared Future for All Life on Earth". During COP15 at the Leaders' Summit, Chinese President Xi Jinping emphasized that, "If we humanity do not fail Nature, Nature will not fail us". Man and Nature need to coexist in harmony, and the international community needs to join hands and start a new journey of high-quality development for humanity. We shall take the development of ecological civilization as our guide to coordinate the relationship between man and Nature. We shall let green transition drive our efforts to facilitate global sustainable development. We shall concentrate on bettering people's well-being to promote social equity and justice. We shall take international law as the basis to uphold a fair and equitable international governance system. The Kunming Declaration has been adopted at COP15, which injected new impetus into global environmental governance. China will establish a Kunming Biodiversity Fund and take the lead by investing 1.5 billion RMB yuan. China has officially designated its first group of national parks, and will put in place a "1+N" policy framework for carbon peak and carbon neutrality.

China always adheres to the concept of ecological civilization to guide development, attaches great importance to biodiversity conservation, and has taken effective actions and achieved remarkable results.

The strategic position of biodiversity has been significantly enhanced. China has incorporated "ecological civilization" into its Constitution, established the China National Committee for Biodiversity

Conservation, implemented "China National Biodiversity Conservation Strategy and Action Plan (2011-2030)", carried out a series of activities of "United Nations Decade on Biodiversity, China in Action", implemented major projects for biodiversity conservation, and strengthened ecological protection and restoration.

The establishment of a system of policies, laws and regulations has been accelerated. China successively issued a number of reform measures related to biodiversity conservation, including "The Overall Plan of Ecological Civilization System Reform", "Opinions on Further Strengthening Biodiversity Conservation", "Suggestions on Delimiting and Strictly Observing the Red Line of Ecological Conservation" and "Suggestions on Improving the Compensation Mechanism for Ecological Conservation". It has integrated biodiversity into plans on economic and social development, ecological conservation and restoration, and territorial and spatial planning. China also formulated the "Biosafety Law" and issued "The Comprehensive Ban on Illegal Wildlife Trade". Since 2021, a 10-year fishing ban policy has been implemented in key waters of the Yangtze River Basin.

The ecological space conservation has been continuously strengthened. China has established the ecological conservation red line system. The preliminarily designated area protected by ecological red lines is not less than 25% of land area, covering key ecological areas, ecologically sensitive and fragile areas and other key ecological areas, and has effectively protected species and their habitats. A system of protected areas, with national parks as the mainstay, has been developed, covering about 11,800 protected areas at various levels and types, accounting for 18% of the land and 4.1% of the sea area.

Fruitful biodiversity conservation achievements have been made. In the past 10 years, China's forest resources have increased by more than 70 million hectares, ranking the first in the world. Desertification has been brought under long-term and large-scale control, wetlands have been effectively protected and restored, and the collection and preservation of genetic resources is among the top rank in the world. 90% of terrestrial ecosystem types, 85% of key wildlife populations and 65% of higher plant communities have been effectively protected. Rare and endangered species have recovered. The total number of wild and breeding crested ibis population has exceeded 6,600, the population of wild giant panda has exceeded 1,800, making giant panda downgraded from "endangered" to "vulnerable" on the global list of species at risk of extinction, and nearly 100 species of wild plants with minimal populations have been under rescue conservation.

Positive progress has been achieved in international exchanges and cooperation. China has firmly supported and practiced multilateralism and earnestly fulfilled its obligations under international conventions on ecological environment. It has made remarkable progress in implementing the "Aichi Targets" on global biodiversity, and its progress towards all 20 "Aichi Biodiversity Targets" is better than the overall situation at international level. China has committed to having CO_2 emissions peak before 2030 and achieving carbon neutrality before

2060. It has proposed and established the BRI International Green Development Coalition (BRIGC) to carry out environmental protection exchanges and cooperation with more than 100 countries. China has steadily promoted the implementation of the Convention on Biological Diversity, and worked with the international community to develop the Post-2020 global biodiversity conservation strategy.

Looking into the future, China will seek development opportunities from the protection of nature and accelerate the construction of an ecological economic system that gives priority to protection and promotes green development. We will improve the technical standard system for biodiversity survey and observation, and the evaluation system for conservation effectiveness. We will continue to carry out major projects to protect biodiversity and improve the quality and stability of ecosystems. We will comprehensively strengthen our capacity for biodiversity conservation and supervision, strictly control the changes in the use of important ecological spaces, and strengthen whole-process supervision over the use of resources related to wildlife. We will raise public awareness and enhance public participation in biodiversity conservation. China will be actively engaged in the process of the Convention on Biological Diversity (CBD) to enhance action to respond to global environmental challenges.

China is ready to work hand in hand with the international community, be responsible for human civilization, respect, conform to and protect nature, and implement the "Transforming Our World: The 2030 Agenda for Sustainable Development", and explore a path of harmonious co-existence between humanity and nature for the sake of the community of all life on Earth, so as to build a clean, healthy, vibrant, prosperous and beautiful world.

Minister of the Ministry of Ecology and
Environment of the People's Republic of China

Contents

目录

《IUCN濒危物种红色保护名录》濒危等级　IUCN Red List of Threatened Species

- **EX** Extinct in the Wild 野外灭绝
- **CR** Critically Endangered 极危
- **EN** Endangered 濒危
- **VU** Vulnerable 易危
- **NT** Near Threatened 近危
- **LC** Least Concern 无危

中国国家重点保护野生动植物的保护等级　Class of Wildlife Under National Priority Protection in China

- **国 I** First-class National Protected Species 国家一级重点保护
- **国 II** Second-class National Protected Species国家二级重点保护

中国生物多样性保护优先区域

Priority Areas of Biodiversity Conservation of China

审图号：GS（2021）6103号

大兴安岭区
小兴安岭区
呼伦贝尔区
松嫩平原区
三江平原区
哈尔滨
长春
长白山区
锡林郭勒草原区
沈阳
山区
呼和浩特
北京
天津
黄渤海区
石家庄
太原
太行山区
济南
黄
海
六盘山-子午岭区
郑州
秦岭区
西安
大别山区
合肥
南京
上海
黄山-怀玉山区
东
海
及
台
湾
海
峡
区
舟山群岛
杭州
巴山区
北段区
武汉
鄱阳湖区
东
海
洞庭湖区
南昌
长沙
武夷山区
赤尾屿
钓鱼岛
陵山区
南岭区
福州
台北
贵阳
区
台湾岛
桂西南山地区
广州
南宁
澳门
香港
东沙群岛
东沙岛
北部湾
南海区
南海区
海口
海南岛
南海区
岛中南部区
南岭区
南宁
桂西南山地区
澳门
香港
广州
台湾岛
河内
南海区
东沙群岛
南海区
海南岛
海南岛中南部区
西沙群岛
中沙群岛
马尼拉
南海区
南
南沙群岛
斯里巴加湾市
南海诸岛

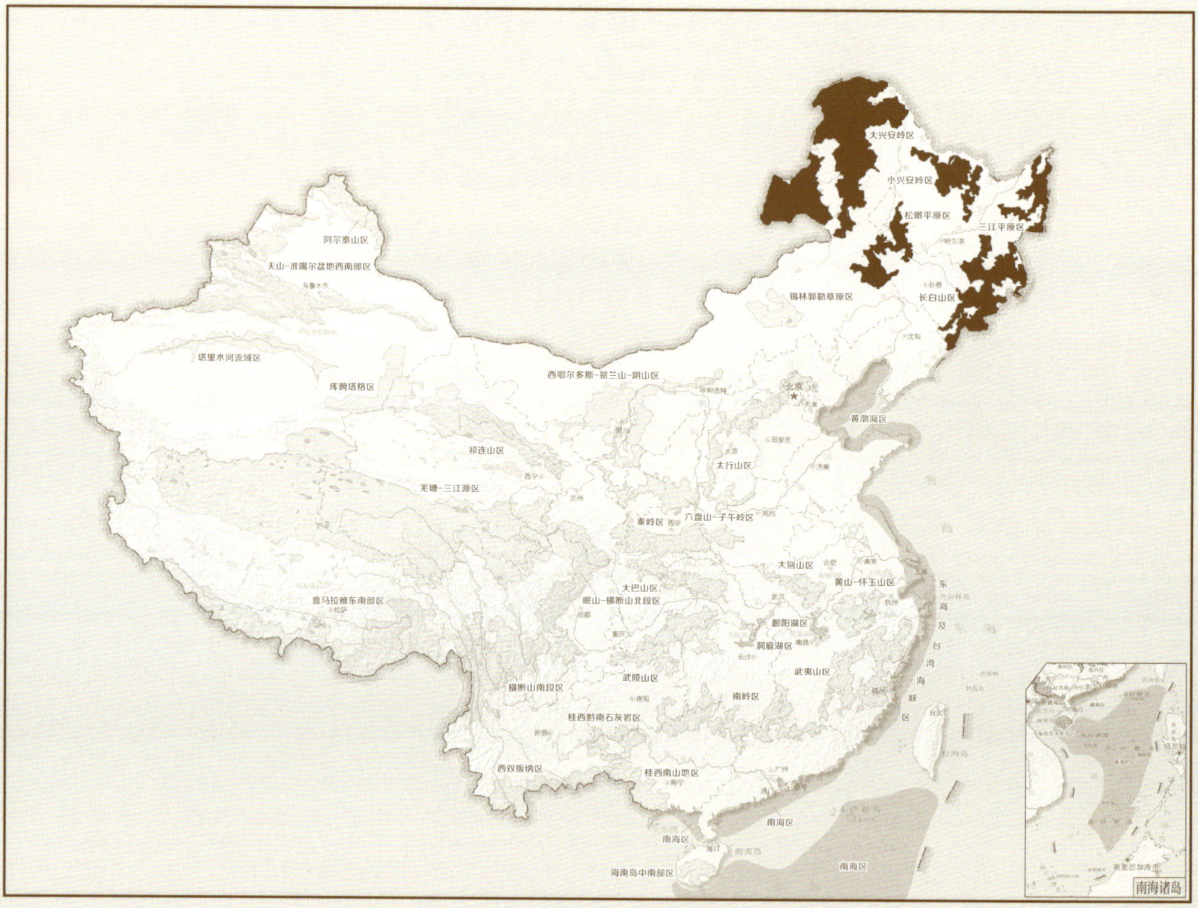
大兴安岭区
小兴安岭区
松嫩平原区
三江平原区
长白山区
锡林郭勒草原区
阿尔泰山区
天山-准噶尔盆地西南部区
塔里木河流域区
库姆塔格区
西鄂尔多斯-贺兰山-阴山区
北京
黄渤海区
祁连山区
太行山区
羌塘-三江源区
秦岭区
六盘山-子午岭区
大别山区
黄山-怀玉山区
大巴山区
喜马拉雅东南部区
岷山-横断山北段区
鄱阳湖区
洞庭湖区
武夷山区
横断山南段区
武陵山区
南岭区
桂西黔南石灰岩区
西双版纳区
桂西南山地区
南海区
海南岛中南部区
东海及台湾海峡区
南海诸岛

Hilly Plain Region of Northeast China

东北山地平原区

东北山地平原区位于中国东北部，主要包括内蒙古自治区东北部、黑龙江省和吉林省等 3 个省（自治区）的 6 个生物多样性保护优先区域，分别为大兴安岭区、小兴安岭区、呼伦贝尔区、三江平原区、长白山区以及松嫩平原区，总面积 380 550 km^2。主要保护对象为湿地生态系统、森林生态系统以及东北虎、丹顶鹤、红松等珍稀濒危野生动植物。保护优先区内建有国家级自然保护区 54 处，总面积 67 638.17 km^2。区内还建有国家公园体制试点 1 处，即东北虎豹国家公园。

Hilly Plain Region of Northeast China majorly consists of 6 priority areas of biodiversity conservation spread across the provinces (autonomous regions) of Heilongjiang, Jilin, and Inner Mongolia, namely Daxing'anling Mountain Area, Xiaoxing'anling Mountain Area, Hulunbeir Area, Three River Plain, Changbaishan Mountain Area, and Songnen Plain, spanning 380,550 km^2 in total. Main subjects of conservation include wetland ecosystems, forest ecosystems, and rare and endangered species, such as *Panthera tigris altaica, Grus monacha* and *Pinus koraiensis*. 54 national nature reserves are located in these priority areas, covering a total area of 67,638.17 km^2. There is, in addition, one of the 10 pilot National Parks, namely the Northeast China Tiger and Leopard National Park.

湿地生态系统

Wetlands Ecosystem

湿地生态系统是东北山地平原区具有代表性的生态系统类型之一，包括河流湿地、沼泽湿地、森林湿地等多种类型。该地区已建成向海、兴凯湖、南瓮河等多个以该生态系统为主要保护对象的国家级自然保护区。

Wetland ecosystems are considered representative ecosystems in Hilly Plain Region of Northeast China, including riverine wetlands, marshes, and swamps. In response to wetlands conservation, multiple national nature reserves, such as Xianghai National Nature Reserve, Xingkai Lake National Nature Reserve and Nanwenghe National Nature Reserve, have been established.

霍超 / 摄　黑龙江南瓮河国家级自然保护区

寒温带针叶林生态系统

Cold-Temperate Coniferous Forest Ecosystem

寒温带针叶林生态系统也是东北山地平原区具有代表性的生态系统类型之一，目前已建立绰纳河、呼中等多个以该生态系统为主要保护对象的国家级自然保护区。

The cold-temperate coniferous forest ecosystem is also one of the representative ecosystem types in the Hilly Plain Region of Northeast China. At present, many national nature reserves with this ecosystem as main protection object have been established, such as Chuonahe and Huzhong.

刘亮 / 摄　黑龙江南瓮河国家级自然保护区

东北虎 国I EN

Amur Tiger
Panthera tigris altaica

东北虎是虎亚种中体型最大的一种，在中国分布于吉林和黑龙江等地。2003 年，吉林首次拍到野生东北虎的照片，证明了中国依然有野生东北虎的存在。作为全球“老虎生存”计划 13 个成员国之一，中国政府出台了《中国野生虎恢复计划》。在东北虎的分布区，现已建立珲春东北虎、老爷岭东北虎等多个国家级自然保护区，并于 2015 年启动东北虎豹国家公园体制试点。近年来，监测数据显示，中国分布的东北虎至少有 55 只。随着保护力度不断加强，此数字还将继续增长。

Amur Tiger is physically the largest amongst all tiger subspecies, distributed across Jilin, Heilongjiang and adjoining regions in China. A wild Amur Tiger was captured by camera for the first time in 2003 in Jilin, proving the possibility of restoring this species' population in China. As one of the 13 member countries of Tigers Alive Initiative of World Wildlife Fund, China has launched China Tiger Recovery Program and established many national nature reserves, such as Hunchun Amur Tiger National Nature Reserve and Laoyeling Amur Tiger National Nature Reserve. In 2015, the Northeast China Tiger and Leopard National Park pilot program was launched. In recent years, the population of Amur Tiger in China has increased to at least 55. This number will continue to grow as conservation efforts intensify.

吉林省野生动植物保护协会 / 供图　吉林珲春东北虎国家级自然保护区

吉林省野生动植物保护协会 / 供图　吉林珲春东北虎国家级自然保护区

白头鹤 国I VU

Hooded Crane
Grus monacha

20 世纪 90 年代初，中国首次在黑龙江发现白头鹤的繁殖巢，成为继俄罗斯之后第二个发现白头鹤繁殖巢的国家，后又发现多个白头鹤繁殖巢，确定了白头鹤在中国的多个繁殖种群，并建立了大沾河湿地和新青白头鹤两个以该物种及其栖息地为重点保护对象的国家级自然保护区。中国学者在白头鹤保护与研究中的突出贡献也获得了国际认可。近二十年，白头鹤的全球种群数量增长了约 70%，保护成果显著。

In early 1990s, a nest of hooded crane was found for the first time in Heilong Jiang, China, making the latter the first country with discovered hooded crane nest in the world besides Russia; more were spotted after further researches, leading to the confirmation of multiple breeding populations, and the establishment of 2 national reserves, Dazhan River Wetland and Xinqing, respectively, specifically for hooded crane conservation. The outstanding contribution that Chinese academics made in the field of hooked crane study is internationally acknowledged. During the past 2 decades, the bird's population grew by a remarkable proportion of 70%.

郭玉民 / 摄

郭玉民 / 摄

郭玉民 / 摄

丹顶鹤 国I EN

Red-crowned Crane

Grus japonensis

丹顶鹤主要繁殖于芦苇沼泽湿地，中国东北地区是其重要的繁殖地，该地区已建立向海、扎龙、兴凯湖等多个以该物种及其栖息地为主要保护对象的国家级自然保护区。在丹顶鹤的繁殖地，退耕还湿、湿地引水、加强巡护、科普宣教等多种保护措施并行，保护丹顶鹤的野生种群及其栖息地。同时，丹顶鹤的人工繁育技术也已日渐成熟。目前主要通过保护野生种群以及野外放归人工繁育种群等方式帮助提升丹顶鹤的种群数量。

Northeast China is a crucial breeding ground for red-crowned cranes, where the birds nest in reed marshes. Many national nature reserves, such as Xianghai, Zhalong, and Xingkai Lake, have been established in their breeding areas. Meanwhile, various protection measures, such as wetlands rehabilitation from agriculture, re-saturation of wetlands, enforced patrols, and public education, have been implemented to protect wild populations and habitats of the cranes. In addition, captive breeding of this species has gradually improved. The number of red-crowned cranes has been increased by such means as protecting wild populations and releasing birds bred in captivity into the wild.

赵俊 / 摄

红松 国II VU

Pinus koraiensis

红松产于中国东北长白山区、吉林山区以及小兴安岭等地，是东北林区的主要森林树种之一，原始红松林也是东北虎重要的栖息地。中国已建立了凉水、小北湖等多个国家级自然保护区，保护天然红松林资源，其中，位于黑龙江省伊春的丰林国家级自然保护区保存着亚洲最完整、面积最大的原始红松林，被纳入联合国教科文组织世界生物圈网络。

Pinus koraiensis, a pine tree species native to the Changbai Mountains, mountainous areas in Jilin, and the Xiaoxing'anling Mountains, is one of the major species in the forests of Northeast China. Virgin *Pinus koraiensis* forests are important habitats for Amur tigers. China has established several national nature reserves, such as Liangshui National Nature Reserve and Xiaobeihu National Nature Reserve, to protect natural *Pinus koraiensis* forests. Among these reserves, Fenglin National Nature Reserve in Yichun, Heilongjiang Province, which has the most complete and largest virgin *Pinus koraiensis* forest in Asia, has been involved in UNESCO's World Biosphere Reserve Network.

陈建伟 自然影像中国 / 摄

陈建伟 自然影像中国 / 摄

大兴安岭区
小兴安岭区
呼伦贝尔区
松嫩平原区
三江平原区
锡林郭勒草原区
长白山区
阿尔泰山区
天山-准噶尔盆地西南部区
塔里木河流域区
库姆塔格区
西鄂尔多斯-贺兰山-阴山区
祁连山区
羌塘-三江源区
黄渤海区
太行山区
秦岭区
六盘山-子午岭区
大别山区
黄山-怀玉山区
大巴山区
岷山-横断山北段区
喜马拉雅东南部区
鄱阳湖区
洞庭湖区
武夷山区
横断山南段区
武陵山区
南岭区
桂西黔南石灰岩区
西双版纳区
桂西南山地区
东海及台湾海峡区
南海区
海南岛中南部区
南海诸岛

Desert Region of Inner Mongolia-Xinjiang Plateau

蒙新高原荒漠区

蒙新高原荒漠区位于中国西北部，包括新疆维吾尔自治区、内蒙古自治区、甘肃省、青海省、宁夏回族自治区 5 个省（自治区）的 7 个生物多样性保护优先区域，这些区域分别为阿尔泰山区、天山一准噶尔盆地西南缘区、塔里木河流域区、祁连山区、库姆塔格区、西鄂尔多斯一贺兰山一阴山区以及锡林郭勒草原区，总面积 551 475 km^2。主要保护对象为荒漠生态系统以及新疆野苹果、雪豹、普氏野马等珍稀濒危野生动植物。保护优先区内建有国家级自然保护区 30 处，总面积 151 702.37 km^2。区内还建有国家公园体制试点 1 处，即祁连山国家公园。

Desert Region of Inner Mongolia-Xinjiang Plateau is located in northwestern China, primarily including 7 priority areas of biodiversity conservation in 5 provinces (autonomous regions) of Xinjiang, Inner Mongolia, Gansu, Qinghai and Ningxia, namely Altay Mountains Area, the Southwestern section of Tianshan-Jungar Basin, Tarim River Basin Area, Qilian Mountain Area, Kumtag Area, West Erdos-Helan Mountain-Yinshan Mountain Area and Xilinguole Pasture Area, with a total area of 551,475 km^2. The main protected objects are desert ecosystems, and rare and endangered species such as *Malus sieversii*, *Panthera uncia* and *Equus ferus*. There are 30 national nature reserves in the priority areas, with a total area of 151,702.37 km^2. There is also a pilot national park in the area, namely Qilian Mountain National Park.

荒漠生态系统

Desert Ecosystem

荒漠生态系统是蒙新高原荒漠区具有代表性的生态系统类型之一，现已建立安西极旱荒漠、民勤连古城、哈巴湖等多个以这一生态系统为主要保护对象的国家级自然保护区。

The desert ecosystem is one of the representative ecosystem types in the desert area of the Desert Region of Inner Mongolia-Xinjiang Plateau. Several national nature reserves, such as Anxi Extremely Dry Desert, Minqin Liangucheng, Haba Lake, etc. have been established with this ecosystem as main protected object.

成勇 / 摄 新疆阿尔金山国家级自然保护区

雪豹 国Ⅰ VU

Snow Leopard
Panthera uncia

中国是雪豹的主要分布国。据估测，现存雪豹栖息地中 60% 的区域位于中国境内。为保护这一高原旗舰物种，中国政府制定了《中国雪豹保护行动计划》，初步建立了以自然保护区为主体的保护管理体系，不断完善野生动物损害补偿体系。山水自然、陆桥生态中心等民间机构也积极参与了雪豹的研究与保护工作。2018 年，19 家科研和自然保护机构联合发布了《中国雪豹调查和保护状况报告 2018》。雪豹 IUCN 濒危等级已由濒危（EN）降至易危（VU）。

Snow leopards are mostly distributed in China, with an estimated 60% of its existing habitats located in the country. In order to protect this iconic plateau species, the Chinese government has formulated Snow Leopard Conservation Action Plan for China, preliminarily established a protection and management system centered around nature reserves, and has improved the compensation system against economic loss caused by wildlife. Civil organizations also eagerly participated in the research and conservation of snow leopards. In 2018, 19 scientific research and nature conservation institutions jointly released *2018 Report on the Investigation and Protection of Snow Leopards of China*. Snow leopard has now been downgraded from Endangered (EN) to VU (Vulnerable) by IUCN Red List of Threatened Species.

陆桥生态中心 祁连山国家公园甘肃省管理局酒泉分局 / 供图

山水自然保护中心 / 供图

山水自然保护中心 / 供图

次丁 野性中国 / 供图

普氏野马 国Ⅰ EN

Przewalski's Horse
Equus ferus

普氏野马，在中国原分布于新疆北部、甘肃、内蒙古一带。曾一度全球野外灭绝。中国于 1986 年在新疆建立了亚洲最大的普氏野马繁育科研中心，开始普氏野马人工繁育工作。1990 年，首次将圈养改为大面积放养，2001 年将第一批 27 匹普氏野马正式放归野外。如今，新疆卡拉麦里有蹄类、安西极旱荒漠及敦煌西湖国家级自然保护区均已开展普氏野马的人工繁育与野外放归工作。目前，中国的普氏野马种群数量已突破 600 匹，占世界普氏野马总数的近 1/3。2015 年，普氏野马 IUCN 濒危等级被下调至濒危（EN）。

Przewalski's horse in China originally found in northern Xinjiang, Gansu and Inner Mongolia, was once considered extinct in the wild. In 1986, China established Asia's largest research center for the captive breeding of Przewalski's horses in Xinjiang. In 1990, captive breeding efforts were replaced by large-scale grazing for the first time. In 2001, the first 27 horses were permanently released into the wild. Currently, captive breeding and releasing of Przewalski's horses are carried out in Mountain Kalamaili Ungulate Nature Reserve of Xinjiang, Anxi Extremely Arid Desert National Nature Reserve and Dunhuang West Lake National Nature Reserve. So far, the number of Przewalski's horses in China has exceeded 600, accounting for nearly one third of the world's total. In 2015, Przewalski's horse was downgraded to Endangered (EN) in the IUCN Red List of Threatened Species.

陈建伟 自然影像中国 / 摄

徐健 / 摄

陈建伟 自然影像中国 / 摄

陈建伟 自然影像中国 / 摄

野骆驼　国I　CR

Bactrian Camel
Camelus ferus

野骆驼是世界上仅存的真驼属野生种。全球种群数量不足 1 000 峰，中国大约有 650 峰，主要分布于库姆塔格沙漠及周边地区、塔克拉玛干沙漠等地。中国已建立罗布泊野骆驼、安南坝野骆驼及敦煌西湖 3 个国家级自然保护区，重点保护该物种及其栖息地，并开展该物种的人工繁育工作。甘肃濒危动物研究中心建立了唯一进行饲养繁育的野骆驼种群。2012 年，在敦煌西湖国家级自然保护区，4 峰个体被野放，到 2019 年已繁育增加到 11 峰。

Bactrian camel is the world's only existing wild species of genus *Camelus*. There are less than 1,000 Bactrian camels living in the world, including about 650 in China, mainly found in Kumtag Desert and beyond, Taklimakan Desert and other areas. China has established three national nature reserves, namely Lop Nur Wild Camel National Nature Reserve, Annanba Wild Camel National Nature Reserve and Dunhuang West Lake National Nature Reserve, to protect this species and its habitats, also working to breed it in captivity. Gansu Endangered Animals Research Center owns the only breeding population of captive Bactrian camels. In 2012, 4 camels were released into the wild in Dunhuang West Lake National Nature Reserve, whose number has grown to 11 by 2019 through breeding.

徐永春 自然影像中国 / 摄

于凤琴 / 摄

陈建伟 自然影像中国 / 摄

伊犁鼠兔 国II EN

Ili Pika
Ochotona iliensis

伊犁鼠兔是中国特有种，仅分布在新疆天山。1983 年被中国学者首次发现，1986 年被正式命名。在之后的几十年里，该物种的命名者持续开展该物种的调查研究与保护工作。2014 年通过红外相机获得该物种影像资料，让这一物种再次进入大众视野，获得广泛关注。同年，天山一号冰川保护区在伊犁鼠兔的主要分布地建立。随着保护和监测力度加强，这一物种又被多次记录到，越来越多的志愿者加入到该物种的保护工作中。2021 年，新调整的《国家重点保护野生动物名录》公布，该物种被列为国家二级重点保护野生动物。

Ili pika is an endemic species in China, only found in the Tianshan Mountains of Xinjiang. It was first discovered in 1983 and officially named in 1986 by Chinese scholars. In the following decades, the scholars who named this species continued with surveys, researches and conservation of this species. In 2014, video footages captured by infrared cameras brought this species back into public attention. In the same year, Tianshan No. 1 Glacier Reserve was established within their main habitat. With more conservation and monitoring efforts devoted, many more footages of the species have been collected. More volunteers have also joined to work on the conservation of this species. In 2021, the newly adjusted “List of National Key Protected Wildlife" was published, and the species was listed as a Class II.

李维东 / 摄

李维东 / 摄

李维东 / 摄

遗鸥 国I VU

Relict Gull
Larus relictus

遗鸥的最重要繁殖地位于中国。20 世纪 90 年代初，中国科学家在内蒙古鄂尔多斯发现了遗鸥的繁殖群，之后在该地区建立了鄂尔多斯遗鸥国家级自然保护区。2002 年，该保护区被列为全球第 1 148 号国际重要湿地。2000 年，陕西省红碱淖湿地首次记录到遗鸥，之后遗鸥在该地区的繁殖种群逐年增多，该保护区现已成为世界上最大的遗鸥繁殖地。随着遗鸥分布地生态补水、湖心岛修复等保护措施的实施，遗鸥的全球种群数量逐步上升。

The most important breeding ground for relict gulls lies in China. In early 1990’s, Chinese scientists discovered breeding communities of relict gulls in Erdos, Inner Mongolia, and then established Erdos Relict Gull National Nature Reserve, which was later listed as Wetland of International Importance No. 1,148 in 2002. In 2000, relict gulls were first recorded in Hongjiannao Wetland of Shaanxi Province. Thereafter, their breeding population sees a continuous growth each year. Today, Hongjiannao National Nature Reserve of Shaanxi Province has become the world’s largest relict gull breeding ground. With reinforced conservation measures, such as ecological water replenishment and restoration of lake islands, the global population of relict gulls has gradually increased.

陈建伟 自然影像中国 / 摄

陈建伟 自然影像中国 / 摄

陈建伟 自然影像中国 / 摄

新疆野苹果 国II VU

Malus sieversii

新疆野苹果在中国仅分布在新疆西部，是重要的种质资源，是世界野苹果基因库的重要组成部分，也是现代栽培苹果的祖先。为使该物种得到有效保护，中国政府已在其分布区建立巴尔鲁克山国家级自然保护区。2016 年，中国科学研究院启动“天山北坡退化野果林生态保育与健康调控技术”项目，保护以新疆野苹果为主要代表的物种。

Malus sieversii, of which west Xinjiang is its only habitat in China, is an important cultivar resource, a significant part of the world's wild apple gene pool, and the ancestor of modern cultivated apples (*Malus domestica*). Baerluke Mountain National Nature Reserve has been established in its distributed area to effectively protect the species. In 2016, Chinese Academy of Sciences launched a project known as "Ecological Conservation and Health Control Technologies of Degraded Wild Fruit Forests on the Northern Slope of the Tianshan Mountains" to protect *Malus sieversii* and other relevant species.

崔大方 / 摄

刘冰 / 摄

刘冰 / 摄

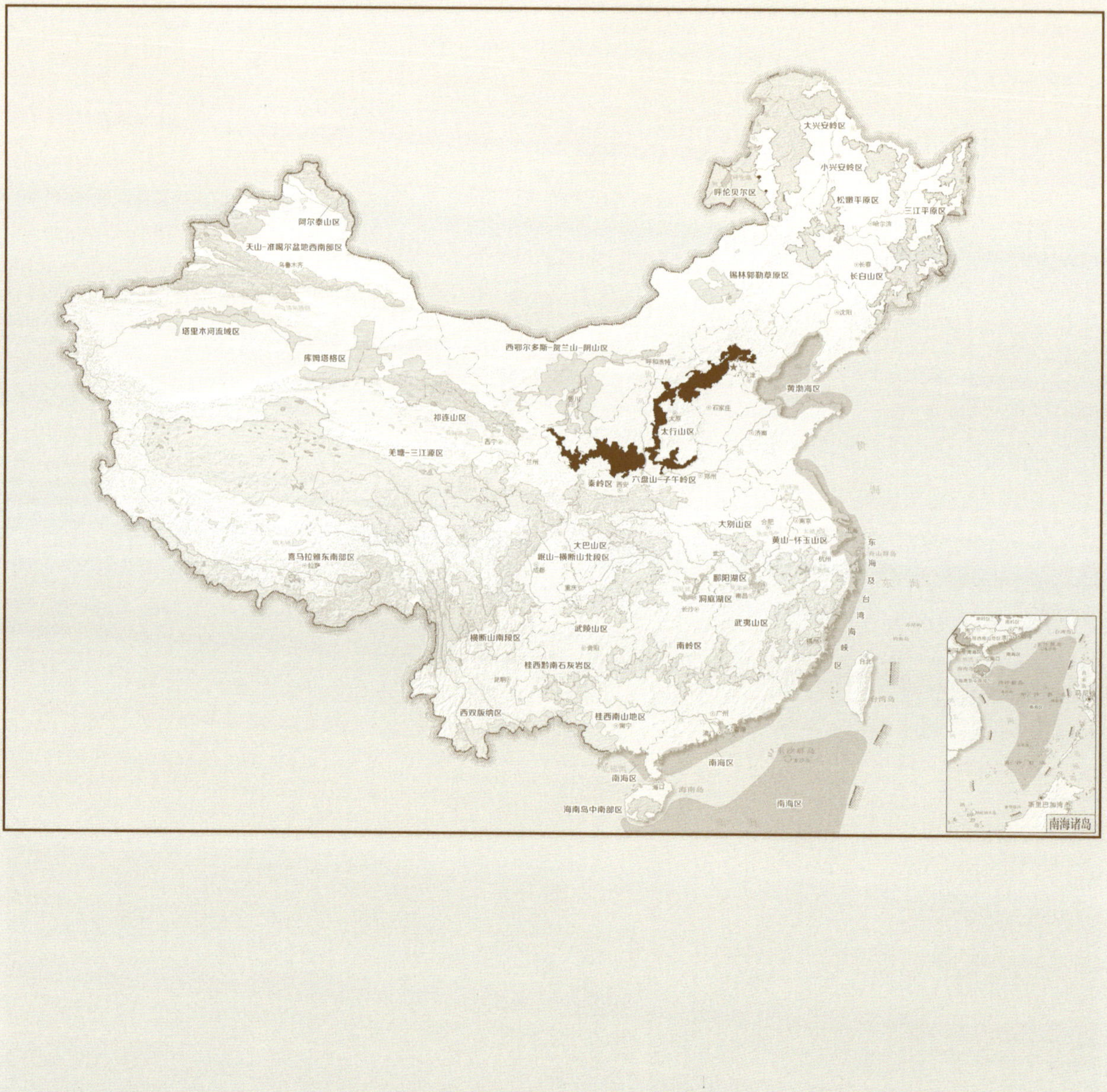
大兴安岭区
呼伦贝尔区
小兴安岭区
松嫩平原区
三江平原区
阿尔泰山区
天山-准噶尔盆地西南部区
锡林郭勒草原区
长白山区
塔里木河流域区
库姆塔格区
西鄂尔多斯-贺兰山-阴山区
黄渤海区
祁连山区
太行山区
羌塘-三江源区
秦岭区
六盘山-子午岭区
大别山区
黄山-怀玉山区
大巴山区
东海及台湾海峡区
喜马拉雅东南部区
岷山-横断山北段区
鄱阳湖区
洞庭湖区
武陵山区
武夷山区
横断山南段区
南岭区
桂西黔南石灰岩区
西双版纳区
桂西南山地区
南海区
南海区
海南岛中南部区
南海区
南海诸岛

Loess Plateau Region and North China Plain

华北平原黄土高原区

华北平原黄土高原区位于中国中东部，主要包括陕西省中部、河北省西部和山西省等 8 个省（市）的 2 个生物多样性保护优先区域，分别为六盘山—子午岭区和太行山区，总面积 105 864 km^2。主要保护对象为暖温带落叶阔叶林生态系统以及华北豹、褐马鸡、太行花等珍稀濒危野生动植物。保护优先区内建有国家级自然保护区 23 处，总面积 5 018.82 km^2。

Loess Plateau Region and North China Plain is located in central-eastern China, mainly consisting of 2 priority areas of biodiversity conservation, respectively Liupanshan-Ziwuling Mountain Area and Taihang Mountain Area, covering 8 provinces (municipality) including central Shaanxi, western Hebei, and Shanxi, with a total area of 105,864 km^2. The main protected subjects are the warm-temperate deciduous broad-leaved forest ecosystem, rare and endangered species such as *Panthera pardus japonensis*, *Crossoptilon mantchuricum* and *Taihangia*. 23 national nature reserves are located in the priority protection areas, with a total area of 5,018.82 km^2.

暖温带落叶阔叶林生态系统

Warm-Temperate Deciduous Broad-Leaved Forest Ecosystem

暖温带落叶阔叶林生态系统是华北平原黄土高原区具有代表性的生态系统类型之一。已建立太统—崆峒山、灵空山等多个以该生态系统为主要保护对象的国家级自然保护区。

The warm-temperate deciduous broad-leaved forest ecosystem is one of the representative ecosystem types in the Loess Plateau Region and North China Plain. Taitong-Kongtong Mountain, Lingkong Mountain and many other national nature reserves have been established to mainly protect this ecosystem.

杨孝 / 摄

华北豹 国I VU

North China Leopard
Panthera pardus japonensis

华北豹是中国特有的豹亚种，分布于山西、陕西、宁夏、内蒙古、甘肃、河北等地。1978 年起，中国在华北地区启动三北防护林、天然林保护等大规模生态工程，同时陆续建立了多个自然保护区，使华北豹及其猎物的栖息地得到保护。近二十年，政府机构、科研院校及民间机构不断加大对华北豹的监测与保护力度，逐渐摸清了华北豹的主要分布地及种群数量。

North China leopard is a leopard subspecies endemic to China, found in Shanxi, Shaanxi, Ningxia, Inner Mongolia, Gansu, Hebei and other areas. Beginning from 1978, China has implemented large-scale ecological projects in the north, such as the Three-North Shelter Forest Program and the Natural Forest Protection Program. Simultaneously, multiple nature reserves have been established successively, ensuring that the habitats and preys of North China leopards are well protected. Over the past two decades, governmental agencies, academic institutions, and NGOs have continuously increased their efforts in monitoring and protecting North Chinese leopards, and gradually revealed the primary range and population size of them.

猫盟 CFCA/ 供图

猫盟 CFCA/ 供图

周哲峰 / 摄

褐马鸡 国I CU

Brown-eared Pheasant
Crossoptilon mantchuricum

褐马鸡是中国特有种，主要栖息在山区森林地带，目前仅分布于山西、陕西、河北、北京等地，是山西省的省鸟。1989 年，中国颁布《野生动物保护法》，使大规模捕杀褐马鸡的现象得到遏制。目前，褐马鸡分布地已建有韩城黄龙山褐马鸡、延安黄龙山褐马鸡、芦芽山等多个以该物种及其栖息地为主要保护对象的国家级自然保护区。褐马鸡的人工繁育工作也取得了一定的进展。

Endemic to China, brown-eared pheasant mainly inhabits forested mountains, and is now restricted in Shanxi, Shaanxi, Hebei, Beijing and other areas. It is the provincial bird of Shanxi. In 1989, China issued Wild Animal Protection Law, restraining the large-scale hunting of brown-eared pheasants. Multiple national nature reserves have now been established in their distribution range, e. g. Hancheng Huanglongshan Brown-Eared Pheasant National Nature Reserve, Yan'an Huanglongshan Brown-Eared Pheasants National Nature Reserve, and Luyashan National Nature Reserve, aiming to protect this species and its habitats. There has also been progress in captive breeding efforts.

冯江 自然影像中国 / 摄

冯江 自然影像中国 / 摄

冯江 自然影像中国 / 摄

东方白鹳 国I EN

Oriental Stork
Ciconia boyciana

东方白鹳曾被认为是白鹳的一个亚种，后被确认为独立种，因面临食物资源少、繁殖巢址缺乏等问题，种群数量较少。为解决这些问题，政府及民间保护机构在东方白鹳的适宜栖息地开展适当人工投食，采取架设人工巢等保护措施。随着各地保护力度的加强，东方白鹳已不局限于仅在中国的东北地区繁殖，河北、山东、江苏、江西等多地均发现了东方白鹳的繁殖种群，其种群数量持续增长。

Ciconia boyciana, or oriental stork, was once regarded as a subspecies of *Ciconia ciconia* and later identified as independent species. Due to such problems as lack of food resources and of nesting sites, its population number was modestly low. To solve these problems, local governments and non-governmental conservation organizations have taken measures like appropriately providing food sources and building artificial nests in suitable habitats. With reinforced local conservation efforts, *Ciconia boyciana* breeds now not only in Northeast China, but also in Hebei, Shandong, Jiangsu, Jiangxi and other areas, leading to a constant population growth.

丁洪安 / 摄

丁洪安 / 摄

丁洪安 / 摄

贾西山 / 摄

太行花属

Taihangia

太行花属是中国特有属，仅包含 1 个种和 1 个变种，即太行花和缘毛太行花，前者仅分布在河南，后者仅分布在河北。2018 年，河北首次发现太行花群落，山西则首次发现缘毛太行花群落，规模均超过万株。这两次发现填补了太行花属植物分布上的空白。在缘毛太行花的分布地已建立青崖寨国家级自然保护区对该物种及其栖息地进行保护，缘毛太行花的迁地保护和引种工作也已开展。

China's endemic plant genus *Taihangia* contains only one species and one variety, with *T. rupestris* only found in Henan, and *T. rupestris* var. *ciliata* only found in Hebei. In 2018, a *T. rupestris* community in Hebei and a *T. rupestris* var. *ciliata* community in Shanxi were first discovered, each of which were made up of more than ten thousand individuals. The discoveries filled the gap in our knowledge of the geographic range of *Taihangia*. In the natural range of *T. rupestris* var. *ciliata*, Qingyazhai National Nature Reserve has been established to protect the flower and its habitats. The ex-situ conservation and introduction of *T. rupestris* var. *ciliata* is also underway.

太行花　刘冰 / 摄

缘毛太行花　陈建伟 自然影像中国 / 摄

太行花　刘冰 / 摄

大兴安岭区
小兴安岭区
呼伦贝尔区
松嫩平原区
三江平原区
阿尔泰山区
天山-准噶尔盆地西南部区
锡林郭勒草原区
长白山区
塔里木河流域区
库姆塔格区
西鄂尔多斯-贺兰山-阴山区
北京
黄渤海区
祁连山区
太行山区
羌塘-三江源区
秦岭区
六盘山-子午岭区
大别山区
黄山-怀玉山区
大巴山区
岷山-横断山北段区
喜马拉雅东南部区
鄱阳湖区
洞庭湖区
东海及台湾海峡区
武陵山区
武夷山区
横断山南段区
南岭区
桂西黔南石灰岩区
西双版纳区
桂西南山地区
南海区
海南岛中南部区
南海诸岛

Alpine Region of Qinghai-Tibetan Plateau

青藏高原高寒区

青藏高原高寒区位于我国西部，主要包括青海省南部和西部及西藏自治区北部和南部等 5 个省（自治区）的 2 个生物多样性保护优先区域，分别为三江源—羌塘区和喜马拉雅山东南区，总面积 979 328 km^2。主要保护对象为高原高寒生态系统以及普氏原羚、黑颈鹤等珍稀濒危野生动植物。保护优先区内建有国家级自然保护区 12 处，总面积 601 933.97 km^2。区内建有国家公园 1 处，即三江源国家公园。

Alpine Region of Qinghai-Tibetan Plateau is located in West China, mainly consisting of 2 priority areas of biodiversity conservation, namely Sanjiangyuan-Qiangtang Area and the Southeast Himalaya Mountain Area, with a total area of 979,328 km^2, in 5 provinces (autonomous regions) including southern and western Qinghai, and northern and southern Tibet. The main protected objects are the alpine meadow ecosystems and rare and endangered species, such as *Procapra przewalskii* and *Grus nigricollis*. There are 12 national nature reserves in the priority protection areas, with a total area of 601,933.97 km^2. There is one national park, the Sanjiangyuan National Park, in the area.

高原高寒草甸生态系统

Alpine Meadow Ecosystem

高原高寒草甸生态系统是青藏高原高寒区具有代表性的生态系统类型之一。该地区已建立可可西里、羌塘等多个以该生态系统为主要保护对象的国家级自然保护区。

The alpine meadow ecosystem is one of the representative ecosystem types of the Alpine Region of Qinghai-Tibetan Plateau. With this ecosystem as the main protected object, a number of national nature reserves including Hoh Xil and Qiangtang have been established.

任红军 / 摄

白颊猕猴 国 II

White-cheeked Macaque
Macaca leucogenys

中国科研工作者在进行西藏野生动物第二次调查（墨脱部分）时，通过野外观察和红外相机记录到了一种猕猴。经过形态学比对，2015 年，最终确认其为猕猴属新物种，并正式命名为白颊猕猴。这是继白头叶猴之后，第二个由中国人发现并命名的灵长类物种。2021 年，新调整的《国家重点保护野生动物名录》公布，该物种被列为国家二级重点保护野生动物。

White-cheeked macaque is a newly-discovered species officially named in 2015 by Chinese researchers, who documented it through infrared cameras and field observation during the second wildlife survey conveyed in Medog County, Tibet. After morphological comparison, it was eventually identified as a new species of the genus *Macaca*. This is the second primate species after white-headed langur discovered and named by Chinese researchers. In 2021, the newly adjusted “List of National Key Protected Wildlife" was published, and the species was listed as a Class II.

李成 / 摄

荒漠猫 国I VU

Chinese Mountain Cat
Felis bieti

荒漠猫是中国唯一的猫科动物特有种，仅分布于中国的青海、四川和甘肃省。北京大学科研团队于 2007 年使用红外相机记录到这一物种，这是荒漠猫科学发现 115 年后首张野外照片。随着政府明令禁止大规模的毒鼠行为等保护措施的实施以及科研力度的加强，在整个青藏高原东缘，从祁连山脉到横断山脉，荒漠猫被不断地目击与记录。2018 年，中国的民间保护机构山水自然在野外调查时，拍摄到了全世界第一组野外繁殖记录。

Chinese mountain cat is the only felid species endemic to China, found only in Qinghai, Sichuan and Gansu. A research team from Peking University documented this species with infrared cameras in 2007, collecting the very first picture of a wild Chinese mountain cat 115 years after the species was discovered. As the government reinforced conservative measures such as the strict prohibition of extensive rat poisoning and scientific researches advanced, across the east edge of Qinghai-Tibet Plateau, from Qilian Mountains to Hengduan Mountains, Chinese mountain cats have been continuously spotted and recorded since. In 2018, during a field investigation, a non-governmental conservation organization, Shan Shui Conservation Center, filmed the reproduction of this species in the wild for the first time in the world.

熊吉吉 / 摄

唐军　西南山地 / 摄

山水自然保护中心 / 供图

普氏原羚 国I EN

Przewalski's Gazelle
Procapra przewalskii

普氏原羚是中国特有种。2001 年，中国将普氏原羚列为 15 类优先保护的物种之一。2004 年，青海湖国家级自然保护区成立普氏原羚保护站，这是中国唯一一个致力于普氏原羚救助、人工驯养繁育的保护站。得益于保护区的建立和全面禁猎，该物种的种群数量从 1994 年的 300 余只增加到如今的 2 700 余只。2008 年，IUCN 将其濒危等级从极度濒危（CR）降为濒危（EN）。

Przewalski's gazelle, an endemic species of China. In 2001, China listed Przewalski's gazelle as one of the 15 preferentially protected species. In 2004, a Przewalski's gazelle conservation station was founded in Qinghai Lake National Nature Reserve, serving as the only conservation station for the rescue, domestication and captive breeding of the species in China. Due to the establishment of reserves and hunting totally banned, the population of this species has increased from 300-odd in 1994 to more than 2,700 today. In 2008, the IUCN downgraded Przewalski's gazelle from Critically Endangered (CR) to Endangered (EN) in the Red List of Threatened Species.

吴秀山 / 摄

徐永春 自然影像中国 / 摄

杨子诚 / 摄

徐永春 自然影像中国 / 摄

藏羚 国I NT

Chiru
Pantholops hodgsonii

藏羚，常被称作藏羚羊，主要分布于中国以羌塘为中心的青藏高原地区。1992—2000 年，为了保护藏羚，野牦牛反盗猎武装队与盗猎分子展开了殊死搏斗，并推动了青海可可西里国家级自然保护区的建立。1999 年，七个国家的代表们在中国西宁讨论并发布了《关于藏羚羊保护及贸易控制的西宁宣言》。在中国政府及民间保护机构的共同努力下，藏羚的种群数量从 19 世纪的约 6 万只恢复到超过 30 万只。2016 年，IUCN 将藏羚的濒危等级由濒危（EN）降至近危（NT）。

Chiru, usually known as Tibetan antelope, is mainly found in Qinghai-Tibet Plateau across the range centered around Qiangtang. From 1992 to 2000, the armed anti-poaching Wild Yak Team fought deathly battles against poachers to save the chirus, and promoted the establishment of Hoh Xil National Nature Reserve in Qinghai Province. In 1999, representatives from seven countries discussed and issued *Xining Declaration on the Conservation and Trade Control of Chirus* in Xining, China. With the mutual efforts of the Chinese government and non-governmental conservation organizations, the population of chirus has recovered from around 60,000 in the 19th century to over 300,000. In 2016, the IUCN downgraded chiru from Endangered (EN) to Near Threatened (NT) in its Red List of Threatened Species.

吴秀山 / 摄

奚志农 野性中国 / 摄

达杰 野性中国 / 摄

黑颈鹤 国I VU

Black-necked Crane
Grus nigricollis

黑颈鹤是世界上唯一一种生活、繁殖在高原的鹤类，在中国主要分布于青藏高原和云贵高原。中国已在青藏高原高寒区建立了隆宝、雅鲁藏布江中游河谷黑颈鹤等多个以该物种及其栖息地为重点保护对象的国家级自然保护区。2015 年，北京林业大学科研团队联合民间保护机构，首次为黑颈鹤搭建人工巢，解决黑颈鹤巢址缺乏的问题，近几年来，已有 27 只幼鹤从人工巢中孵出。黑颈鹤全球种群数量稳步上升。

Black-necked crane is the world's only crane species that lives and breeds on plateaus. In China, it is mainly found in Qinghai-Tibet Plateau and Yunnan-Guizhou Plateau. In alpine regions of Qinghai-Tibet Plateau, China has established several national nature reserves to protect this species and its habitats, such as Longbao National Nature Reserve and Middle Yarlung Zangbo River—the Black-necked Cranes National Nature Reserve. In 2015, to solve the lack of nesting sites, a research team from Beijing Forestry University collaborated with non-governmental conservation organizations to set up artificial nests for black-necked cranes for the first time. In recent years, 27 young cranes have hatched in these artificial nests. The global population of black-necked crane is continuously rising.

唐军 西南山地 / 摄

于凤琴 / 摄

于凤琴 / 摄

大兴安岭区
小兴安岭区
呼伦贝尔区
松嫩平原区
三江平原区
阿尔泰山区
天山-准噶尔盆地西南部区
锡林郭勒草原区
长白山区
塔里木河流域区
库姆塔格区
西鄂尔多斯-贺兰山-阴山区
北京
黄渤海区
祁连山区
太行山区
羌塘-三江源区
秦岭区
六盘山-子午岭区
大别山区
黄山-怀玉山区
大巴山区
岷山-横断山北段区
喜马拉雅东南部区
鄱阳湖区
洞庭湖区
武夷山区
武陵山区
横断山南段区
南岭区
桂西黔南石灰岩区
西双版纳区
桂西南山地区
南海区
海南岛中南部区
南海区
东海及台湾海峡区
南海诸岛

Alpine Canyon Region of Southwest China

西南高山峡谷区

西南高山峡谷区位于我国西南部，主要包括云南省北部和四川省中东部等 4 个省（自治区）的 2 个生物多样性保护优先区域，分别为横断山南段区和岷山—横断山北段区，总面积 216 846km^2。主要保护对象为高山山地森林生态系统以及滇金丝猴、大熊猫、珙桐等珍稀濒危野生动植物。保护优先区内建有国家级自然保护区 31 处，总面积 38 539.12 km^2。区内还建有国家公园体制试点 1 处，即大熊猫国家公园。

Alpine Canyon Region of Southwest China mainly includes 2 priority areas of biodiversity conservation, respectively, the Southern section of Hengduan Mountain, and the Northern section of Minshan-Hengduan Mountain, with a total area of 216,846 km^2, in 4 provinces (autonomous regions) including northern Yunnan and central-eastern Sichuan. The main protected objects are the alpine and mountainous forest ecosystems and rare and endangered wild species, such as *Rhinopithecus bieti*, *Ailuropoda melanoleuca* and *Davidia involucrata*. There are 31 national nature reserves in the priority areas, with a total area of 38,539.12 km^2. A pilot national park, the Giant Panda National Park, is also in this region.

山地森林生态系统

Mountain Forest Ecosystem

山地森林生态系统是西南高山峡谷区具有代表性的生态系统类型之一。已建立白马雪山、贡嘎山等多个以该生态系统为主要保护对象的国家级自然保护区。

Mountain forest ecosystem is one of the representative ecosystem types in the Alpine Canyon Region of Southwest China. A number of national nature reserves, such as Baimaxueshan and Gonggashan, have been established with this kind of ecosystem as the main protected object.

于凤琴 / 摄

滇金丝猴 国I EN

Black Snub-nosed Monkey
Rhinopithecus bieti

滇金丝猴是中国特有种，仅分布于云南省和西藏自治区。在该物种的分布地已建立白马雪山、芒康滇金丝猴国家级自然保护区，用以保护该物种及其栖息地。时隔近 50 年，滇金丝猴于 2007 年在云南云龙天池省级自然保护区被重新发现，该保护区也于 2012 年晋升为国家级自然保护区。2015 年，云南省启动《滇金丝猴云南省保护行动计划（2015—2020 年）》，2017 年，政府机构、科研院校、民间机构联合启动全球首次滇金丝猴全境动态监测。

Black snub-nosed monkey is a species endemic to China, only found in Yunnan and Tibet. Baimaxueshan National Nature Reserve and Mangkang Black Snub-nosed Monkey National Nature Reserve, have been established within its range to protect the species and its habitats. In 2007, after nearly half a century without witness record, black snub-nosed monkey was rediscovered in Yunlong Tianchi Provincial Nature Reserve in Yunnan, which was promoted to national level in 2012. In 2015, Yunnan Province launched *Action Plan of Yunnan Province for the Conservation of Black Snub-Nosed Monkeys (2015-2020)*. In 2017, several government agencies, academic institutions and civil organizations cooperated to launch the world's first dynamic monitoring project of black snub-nosed monkey.

李飚 西南山地 / 摄

于凤琴 / 摄

于凤琴 / 摄

于凤琴 / 摄

高黎贡白眉长臂猿 国I EN

Skywalker Hoolock Gibbon
Hoolock tianxing

高黎贡白眉长臂猿也被称作天行长臂猿，在中国仅分布于云南省，长期被认为是东白眉长臂猿。中国科学家通过比对这一群体与其他东白眉长臂猿的形态学及分子学差异，于 2017 年宣布其独立为种，并命名为高黎贡白眉长臂猿。这也是第一种由中国科学家命名的类人猿。云山保护等民间机构联合当地政府通过开展宣传教育、加强保护地巡逻等方式对该物种及其栖息地进行保护。

Skywalker hoolock gibbon, also known as Gaoligong hoolock gibbon, is only found in Yunnan Province in China, and has long been misidentified as East hoolock gibbon. By comparing morphological and molecular differences between this population and other East hoolock gibbons, Chinese scientists claimed it as an independent species in 2017 and named it *Hoolock tianxing*, making it the first ape species named by Chinese. Civil organizations, such as Cloud Mountain Conservation, in association with the local government, have been implementing education and enhancing patrol efforts within the reserves to protect this species and its habitats.

赵超 / 摄

关翔宇 西南山地 / 摄

董磊 西南山地 / 摄

大熊猫 国I VU

Giant Panda
Ailuropoda melanoleuca

大熊猫是中国特有种，目前仅分布在四川、甘肃和陕西省。20 世纪 90 年代，中国开始实施“中国保护大熊猫及其栖息地工程”。全国第四次大熊猫调查显示，野外大熊猫的数量已从 20 世纪七八十年代的 1 114 只增长到 1 864 只。大熊猫的人工繁育工作也取得了一定成功，如今，人工繁殖种群数量已超过 600 只。为进一步加强对大熊猫及其栖息地的保护，中国建立了大熊猫国家公园体制试点。大熊猫濒危等级已由濒危（EN）降至易危（VU）。

Endemic to China, Giant panda is found only in Sichuan, Gansu and Shaanxi. In 1990s, China launched *China Conservation Program of Giant Pandas and Their Habitats*. According to the fourth nation-wide census of giant panda, the wild population of this species has increased from 1,114 in 1970s and 1980s to 1,864. Captive breeding of giant panda is also considerably gaining: the number of individuals bred in captivity has exceeded 600 today. To further reinforce the conservation of giant panda and its habitat, the pilot program of Giant Panda National Park has been commenced. IUCN downgraded giant panda from Endangered(EN) to Vulnerable(VN) in IUCN Red List of Threatened Species.

冯江 自然影像中国 / 摄

赵纳勋 / 摄

向定乾 / 摄

冯江 自然影像中国 / 摄

白唇鹿 国I VU

White-lipped Deer
Cervus albirostris

白唇鹿是中国特有种，分布于甘肃、青海、西藏、四川、云南等地，栖息在海拔3 000～5 000 m的针叶林和高山草甸中，是现今分布海拔最高的鹿科动物，其分布的南缘在青藏高原高寒区内。在该物种的分布地已建立了四川察青松多白唇鹿等多个国家级自然保护区，以保护该物种及其栖息地。白唇鹿的人工繁殖工作也取得一定进展。

White-lipped deer is a species endemic to China, found in Gansu, Qinghai, Tibet, Sichuan, Yunnan, and other areas. Inhabiting coniferous forests and alpine meadows at an altitude of 3,000 to 5,000 meters, the species is the extant member of Cervidae that dwells at the highest altitude, with the south edge of its range lying in the alpine region of Qinghai-Tibet Plateau. To protect the deer and its habitats, Chaqingsongduo White-lipped Deer National Nature Reserve in Sichuan and several other national nature reserves have been established within its distribution range. Captive breeding of white-lipped deer has also been making progress.

更求曲朋 野性中国 / 摄

达布西力特 / 摄

张强 / 摄

珙桐 国I LC

Davidia involucrata

珙桐是中国特有的孑遗植物，野生种群分布于陕西、湖北、湖南、四川、重庆、贵州和云南等地。在该物种的分布地已建立了多个国家级自然保护区，以保护该物种及其栖息地。其中，位于云南昭通的乌蒙山国家级自然保护区内保存着完整的大面积野生天然珙桐群。2008 年，在四川雅安发现了近 6 600 公顷野生珙桐，是目前中国面积最大的野生珙桐分布区。如今，珙桐也被作为观赏植物引种到多地。

Davidia involucrata is a relict plant species endemic to China, of which the wild populations are found in areas including Shaanxi, Hubei, Hunan, Sichuan, Chongqing, Guizhou, and Yunnan. Several national nature reserves have been established within its natural range to protect the species and its habitat. Located in Zhaotong of Yunnan Province, the Wumengshan National Nature Reserve in particular, is extant with large areas of wild dove trees. In 2008, nearly 6,600 hectares of wild dove trees were discovered in Ya'an, Sichuan Province, making it the largest population of the species in China so far. Today, *Davidia involucrata* has been extensively introduced and cultivated as ornamental plant.

巫嘉伟 西南山地 / 摄

王进 西南山地 / 摄

大兴安岭区
小兴安岭区
呼伦贝尔区
松嫩平原区
三江平原区
长白山区
锡林郭勒草原区
阿尔泰山区
天山-准噶尔盆地西南部区
塔里木河流域区
库姆塔格区
西鄂尔多斯-贺兰山-阴山区
黄渤海区
太行山区
祁连山区
羌塘-三江源区
秦岭区
六盘山-子午岭区
大别山区
黄山-怀玉山区
大巴山区
岷山-横断山北段区
喜马拉雅东南部区
鄱阳湖区
洞庭湖区
武陵山区
武夷山区
横断山南段区
南岭区
桂西黔南石灰岩区
西双版纳区
桂西南山地区
南海区
南海区
海南岛中南部区
南海区
东海及台湾海峡区
南海诸岛

Hilly Region of Central, South and West China

中南西部山地丘陵区

中南西部山地丘陵区位于中国西南部，主要包括陕西省、四川省、湖北省、广西壮族自治区和贵州省等 9 个省（自治区）的 4 个生物多样性保护优先区域，分别为秦岭区、武陵山区、大巴山区和桂西黔南石灰岩区，总面积 200 234 km^2。主要保护对象为亚热带森林生态系统以及黔金丝猴、朱鹮等珍稀濒危野生动植物。保护优先区内建有国家级自然保护区 58 处，总面积 16 849.16 km^2。

Hilly Region of Central, South and West China mainly include 4 priority areas of biodiversity conservation, namely Qinling Mountain Area, Wuling Mountain Area, Daba Mountain Area and the Limestone Area of West Guangxi and South Guizhou, with a total area of 200,234 km^2, in 9 provinces (autonomous regions) including Shaanxi, Sichuan, Hubei, Guangxi and Guizhou. The main protected objects are the subtropical forest ecosystems and rare and endangered species, such as *Rhinopithecus brelichi* and *Nipponia nippon*. There are 58 national nature reserves in the priority areas, with a total area of 16,849.16 km^2.

亚热带森林生态系统

Subtropical Forest Ecosystem

亚热带森林生态系统是中南西部山地丘陵区具有代表性的生态系统类型之一。已建立八大公山、五里坡等多个以该生态系统为主要保护对象的国家级自然保护区。

The subtropical forest ecosystem is one of the representative ecosystem types in the Hilly Regions of Central, South and West China. A number of national nature reserves, such as Badagongshan and Wulipo, have been established with this ecosystem as the main protected object.

税晓洁 / 摄

川金丝猴 国I EN

Golden Snub-nosed Monkey
Rhinopithecus roxellana

川金丝猴是中国特有种，分布于四川、甘肃、湖北和陕西等地。中国建立了分布有(或可能有)川金丝猴的自然保护区 38 个，其中主要以川金丝猴及其栖息地为保护对象的 15 个。2007 年，神农架国家级自然保护区川金丝猴人工补食实验取得成功。2014 年，全国首个关于川金丝猴保护的专题讨论会在四川成都召开，商讨川金丝猴的保护措施。随着栖息地保护的加强，川金丝猴野外种群数量持续增长，川金丝猴的人工繁育工作也取得一定进展。

Golden snub-nosed monkey, endemic to China, is found in Sichuan, Gansu, Hubei and Shaanxi. China has established 38 nature reserves in which golden snub-nosed monkeys inhabit (or possibly inhabit), with 15 of them mainly target the monkey and its habitat as protected objects. In 2007, a trial of food supplement for golden snub-nosed monkeys succeeded in Shennongjia National Nature Reserve. In 2014, the first national symposium on the conservation of golden snub-nosed monkey was held in Chengdu, Sichuan to discuss the conservation measures of the species. With reinforced habitat protection, the population of wild golden snub-nosed monkey continues to grow. Progresses have also been made in captive breeding.

徐永春 自然影像中国 / 摄

徐永春 自然影像中国 / 摄

徐永春 自然影像中国 / 摄

何晓安 西南山地 / 摄

黔金丝猴 国I EN

Grey Snub-nosed Monkey
Rhinopithecus brelichi

黔金丝猴是中国特有种，仅分布在贵州梵净山。在该物种的分布地已建立梵净山国家级自然保护区，并于 1986 年被纳入联合国教科文组织世界生物圈保护区网络。黔金丝猴的人工繁殖技术也取得一定成功，目前，已成功人工繁殖出第三代黔金丝猴。

Grey snub-nosed monkey, a Chinese endemic species, is found only in Fanjing Mountain, Guizhou. Fanjing Mountain National Nature Reserve was established in the distribution range of this species and was involved into UNESCO's World Biosphere Reserve Network in 1986. Progresses have been made in captive breeding techniques of this monkey. So far, three generations of captive grey snub-nosed monkeys have been bred.

陈建伟 自然影像中国 / 摄

陈建伟 自然影像中国 / 摄

丁宽亮 / 摄

朱鹮 国I EN

Asian Crested Ibis
Nipponia nippon

1981 年，中国科学家在陕西洋县发现了一个数量仅为 7 只的朱鹮种群，是当时世界上已知的最后一个野生朱鹮种群，当地政府立即展开了严格的保护工作。1986 年，北京动物园开始尝试朱鹮人工繁育工作，2004 年，陕西启动朱鹮野放工作，均取得成功。多年来，得益于陕西汉中朱鹮等国家级自然保护区的建立，一套完整的朱鹮野外保护体系得以形成，并在陕西、北京、河南、浙江和广东等多地建立了稳定的人工种群和野生种群。如今，中国朱鹮的数量现已超过 6 600 只，野生种群数量超过 4 450 只。IUCN 将朱鹮濒危等级由极度濒危（CR）降至濒危（EN）。

In 1981, Chinese scientists discovered a population of 7 Asian crested ibises in Yangxian, Shaanxi Province, which proved the last known wild population then in the world. The local government immediately took strict measures for conservation. In 1986, Beijing Zoo began the attempt of breeding Asian crested ibises in captivity. In 2004, Shaanxi commenced the releasing of captive-bred individuals back to the wild and succeeded. Thanks to the establishment of Shaanxi Hanzhong Asian Crested Ibis National Nature Reserve and other national nature reserves, over the years, a complete system of field protection for Asian crested ibises has been formed. Stable captive and wild populations have been established in areas such as Shaanxi, Beijing, Henan, Zhejiang, and Guangdong. The number of Asian crested ibises in China has exceeded 6,600, including over 4,450 in the wild. IUCN downgraded Asian crested ibises to Endangered (EN) in IUCN Red List of Threatened Species.

廖国庆 / 摄

赵纳勋 / 摄

张强 / 摄

孙晋强 自然影像中国 / 摄

大鲵 国II CR

Chinese Giant Salamander
Andrias davidianus

大鲵，俗称“娃娃鱼”，因其叫声而得名，是中国特有种，也是体型最大的两栖动物，曾广泛分布于中国中部、南部和西南等地。中国已在大鲵的原产区建立了 47 个以大鲵为主要保护对象或与大鲵相关的自然保护区，如张家界大鲵国家级自然保护区等，此外也建立了一些非原生地大鲵保护区，作为迁地种群保护的种质资源储存库。同时开展大鲵的人工繁育、增殖放流工作。目前，大鲵放流数量累计超过 27 万尾。

Chinese giant salamander (also known as “baby fish” in Chinese, named after its unique call) is an endemic species in China, the largest amphibian in the world, and was once commonly distributed across Central, South and Southwest China. Within the nature range of the species, China has established 47 nature reserves, e. g. the National Giant Salamander Nature Reserve of Zhangjiajie, either focusing on the conservation of or associated with Chinese giant salamanders. Additionally, some non-native reserves for Chinese giant salamanders have also been established as germplasm repositories to protect ex-situ populations. At the same time, captive breeding and releasing of Chinese giant salamanders are also conducted. So far, more than 270,000 salamanders have been released into the wild.

周佳俊 / 摄

陈建伟 自然影像中国 / 摄

周佳俊 / 摄

桫椤科植物是蕨类植物中一个十分独特的类群，该科植物都是树形蕨。中国分布的桫椤科植物包括 2 属 14 种，均为国家二级重点保护野生植物，主要分布在华南和西南地区。目前已建立贵州赤水桫椤、四川画稿溪等多个以桫椤科植物及其栖息地为主要保护对象的国家级自然保护区。

Generally known as scaly tree ferns, family Cyatheaceae represents a quite unique fern group, all members of which are tree ferns. There are 14 species from 2 genera of Cyatheaceae in China, all of which are listed in the Nationally Conserved Wild Plant Catogory II, and mainly found in South and Southwest China. Several national nature reserves for Cyatheaceae and their habitats have been established, such as Chishui Cyatheaceae National Nature Reserve and Huagaoxi National Nature Reserve.

黄耀华 西南山地 / 摄

秦隆 / 摄

陈建伟 自然影像中国 / 摄

阿尔泰山区
天山-准噶尔盆地西南部区
乌鲁木齐
塔里木河流域区
库姆塔格区
祁连山区
西宁
兰州
羌塘-三江源区
喜马拉雅东南部区
拉萨
西鄂尔多斯-贺兰山-阴山区
呼和浩特
银川
大兴安岭区
小兴安岭区
呼伦贝尔区
松嫩平原区
三江平原区
哈尔滨
锡林郭勒草原区
长春
长白山区
沈阳
北京
天津
黄渤海区
石家庄
太原
太行山区
济南
秦岭区
西安
六盘山-子午岭区
郑州
大别山区
合肥
南京
黄山-怀玉山区
大巴山区
岷山-横断山北段区
成都
重庆
武汉
杭州
鄱阳湖区
洞庭湖区
南昌
长沙
武夷山区
武陵山区
贵阳
横断山南段区
南岭区
桂西黔南石灰岩区
昆明
西双版纳区
桂西南山地区
南宁
广州
东
海
及
台
湾
海
峡
区
台北
南海区
南海区
海口
海南岛中南部区
南海区
南海诸岛

Hilly Plain Region of East and Central China

华东华中丘陵平原区

华东华中丘陵平原区位于中国中西部，主要包括江西、安徽、浙江、福建和湖南等9个省（自治区）的6个生物多样性保护优先区域，分别为黄山—怀玉山区、大别山区、武夷山区、南岭区、洞庭湖区和鄱阳湖区，总面积27 735.11 km^2。主要保护对象为湖泊湿地生态系统以及长江江豚、白鹤、百山祖冷杉等珍稀濒危野生动植物。保护优先区内建有国家级自然保护区69处，总面积13 715.75 km^2。

Hilly Plain Region of East and Central China mainly consists of 6 priority areas of biodiversity conservation, namely Huangshan-Huaiyushan Mountain Area, Dabieshan Mountain Area, Wuyishan Mountain Area, Nanling Mountain Area, Dongting Lake Area and Poyang Lake Area, with a total area of 27,735.11 km^2, in 9 provinces (autonomous regions) including Jiangxi, Anhui, Zhejiang, Fujian and Hunan. The main protected objects are the lake and wetland ecosystems and rare and endangered species, e.g. *Neophocaena asiaeorientalis, Leucogeranus leucogeranus* and *Abies beshanzuensis*. There are 69 national nature reserves in the priority areas, with a total area of 13,715.75 km^2.

湖泊湿地生态系统

Lacustrine Ecosystem

湖泊湿地生态系统是华东华中丘陵平原区具有代表性的生态系统类型之一。已建立以洪湖、东洞庭湖、鄱阳湖候鸟等生态系统为主要保护对象的国家级自然保护区。

Lacustrine ecosystem is one of the representative ecosystem types in the Hilly Plains of East China and Central China. Honghu Lake, East Dongting Lake, Poyang Lake, and other national nature reserves have been established with the ecosystem as the main protection target.

何小白 / 摄

长江江豚 国Ⅰ EN

Yangtze Finless Porpoise
Neophocaena asiaeorientalis

长江江豚原为窄脊江豚的一个亚种。2018 年中美科学家研究确认该物种为独立物种，其成为中国长江中下游地区特有的鲸豚类动物。中国确立了“就地保护、迁地保护和人工繁育研究相结合”的长江江豚保护战略。我国首个长江江豚迁地保护种群位于湖北天鹅洲白鱀豚国家级自然保护区内，种群数量已从最初引入的 5 头增长至 80 头，并成为种质资源库。2020 年，首例人工繁殖江豚被成功野放；同年，长江流域开始实施为期 10 年的全面禁渔。2021 年，新调整的《国家重点保护野生动物名录》公布，长江江豚升级为国家一级重点保护野生动物。

Yangtze finless porpoise was previously considered as subspecies of narrow-ridged finless porpoise. In 2018, Chinese and American scientists identified it as an independent species, and it became an endemic cetacean in the middle and lower reaches of the Yangtze River in China. China has established a conservation strategy of combining in-situ protection, ex-situ protection and captive breeding research. China's first ex-situ protected population of Yangtze finless porpoise, located in Tianezhou Baiji National Nature Reserve in Hubei Province, now harbors a population of 80 introduced Yangtze finless porpoises, having developed from the initial 5 individuals and now become a genetic resource repository. In 2020, the first captive-bred Yangtze finless porpoise was released into the wild. In the same year, a 10-year thorough fishing ban covering the entire basin of the Yangtze River went into effect. In 2021, the newly adjusted “List of National Key Protected Wildlife" was published, and the species was listed as a Class I.

余会功 / 摄

余会功 / 摄

麋鹿 国Ⅰ EW

Père David's Deer
Elaphurus davidianus

麋鹿是中国特有种。1985 年，北京南海子从英国乌邦寺重引入 22 头麋鹿，次年，江苏大丰再次引入 39 头麋鹿，并建立了全世界第一个麋鹿自然保护区——大丰麋鹿自然保护区，开启了中国麋鹿保护之路。1997 年，大丰麋鹿自然保护区晋升为国家级自然保护区。1998 年，大丰麋鹿国家级自然保护区开始进行麋鹿野化放归并取得成功。如今，中国已实现圈养麋鹿的野外放归，并成功建立可自我维持的自然种群。现在，中国麋鹿已由 1988 年的近百头发展到超 8 000 头。

Père David's deer is a species endemic to China. In 1985, Beijing Nanhaizi Park reintroduced 22 Père David's deers from Woburn Abbey, England. In the following year, 39 Père David's deer were introduced to Dafeng, Jiangsu, where the world's first nature reserve for it, the Dafeng Milu Nature Reserve, was established, laying the foundation for the conservation of the species. In 1997, Dafeng Milu Nature Reserve was upgraded to a national nature reserve. In 1998, Dafeng Milu National Nature Reserve managed to release Père David's deer into the wild. Today, China has made possible the releasing of captively bred Père David's deers, which have successfully formed self-sustaining natural populations. The number of Père David's deer in China has grown from nearly 100 in 1988 to over 8,000 now.

韦晔 / 摄

于凤琴 / 摄

徐永春 自然影像中国 / 摄

白鹤 国I CR

Siberian Crane
Leucogeranus leucogeranus

白鹤是一种长距离迁徙的大型候鸟，中国是其重要的迁徙停息地和越冬地。白鹤每年有约 7 个月的时间栖息在中国境内，在其入境的首站停歇地，已建立南瓮河等国家级自然保护区。江西鄱阳湖是白鹤最重要的越冬地，每年有约 98% 的全球种群个体在此越冬长达 4~5 个月，在越冬地已建立鄱阳湖南矶湿地、鄱阳湖候鸟等多个以白鹤及其栖息地为主要保护对象的国家级自然保护区。2019 年，白鹤被定为江西省省鸟。

Siberian crane is a large migratory bird that practices long-distance migration, spending about 7 months annually in China. As its important stopover habitat and wintering ground, China has established several national nature reserves, such as Nanwenghe National Nature Reserve, at the crane's first stopover habitat on its way entering the country. Poyang Lake, lying in Jiangxi Province, is one of the most crucial wintering grounds for this species, supporting 98% of individuals from all the world for 4-5 months. Within the wintering ground, Poyang Lake Nanji Wetland National Nature Reserve and Poyang Lake Migratory Birds National Nature Reserve have been established to conserve this species and its habitats. In 2019, Siberian crane was appointed as the provincial bird of Jiangxi.

陈建伟 自然影像中国 / 摄

冯江 自然影像中国 / 摄

冯江 自然影像中国 / 摄

扬子鳄 国I CR

Chinese Alligator
Alligator sinensis

扬子鳄是中国特有种，仅分布于长江中下游地区。为充分保护这一物种及其栖息地，其分布地已建立安徽扬子鳄国家级自然保护区，控制扬子鳄分布地的人为干扰，避免污水和污染物对扬子鳄栖息地的破坏。中国扬子鳄的人工繁殖技术成熟，繁育已具相当规模，种群数量超过万条，并具备了年增2 000条的繁育生产能力；此外，放归野外的扬子鳄已实现自然繁殖。

Chinese alligator is a species endemic to China, only found in the middle and lower reaches of the Yangtze River. As to plenarily conserve this species and its habitats, Anhui Chinese Alligator National Nature Reserve was established in its distribution range to restrain the human interference within and to prevent the damage by sewage and pollutants. As captive breeding techniques develops, the population has been considerably increased; the number has exceeded 10,000, while annual breeding capacity has reached 2,000. The Chinese alligators released into the wild are now capable of reproducing naturally.

薄顺奇 / 摄

徐健 / 摄

谢建国 自然影像中国 / 摄

银杏 国I EN

Ginkgo biloba

银杏是中国特有种，是著名的孑遗植物，野生种群仅分布在浙江西天目山。在该物种的分布地已建立浙江天目山国家级自然保护区，并于 1996 年加入联合国教科文组织“人与生物圈”保护网。银杏的人工栽培繁育技术成熟，已作为行道树、景观树等观赏树种在全国各地广泛种植，国外多个国家也有引种。

Ginkgo biloba is a species endemic to China and a well-known relict plant, of which the wild populations are found only in West Tianmushan, Zhejiang Province, where Zhejiang Tianmushan National Nature Reserve is located. The reserve was involved in UNESCO's Man and the Biosphere Program in 1996. With mature artificial cultivation technologies, *Ginkgo biloba* has now been widely planted as street tree and ornamental tree species all over the country, and are introduced to many foreign countries.

秦隆 / 摄

秦隆 / 摄

陈建伟 自然影像中国 / 摄

百山祖冷杉 国I CR

Abies beshanzuensis

百山祖冷杉是中国特有的孑遗植物，野外种群仅分布在浙江百山祖，1976 年正式命名时仅剩 3 株。其分布地已建立凤阳山—百山祖国家级自然保护区，对该物种及其栖息地进行严格保护。同时，百山祖冷杉人工繁育和迁地保护工作已取得一定成果，1992 年，实生苗移栽成功。2017 年，浙江大学与保护区合作，培育出世界首例百山祖冷杉胚胎，到 2020 年，百山祖冷杉自然授粉的种子在原地传播，已经长出 400 余株幼苗。

Abies beshanzuensis is a relict plant endemic to China, with wild populations found only in Baishanzu, Zhejiang Province. This species had merely three remaining individuals when it was officially named in 1976, leading to the establishment of Fengyangshan-Baishanzu National Nature Reserve to strictly protect this species and its habitats. At the same time, cultivation and ex-situ conservation of *Abies beshanzuensis* have seen positive outcomes. In 1992, seedlings were successfully transplanted and, in 2017, Zhejiang University collaborated with the Nature Reserve to cultivate the world's first *Abies beshanzuensis* embryo. In 2020, naturally pollinated seeds of *Abies beshanzuensis* were spread in situ. To date, more than 400 seedlings have grown.

陈建伟 自然影像中国 / 摄

钟鑫 / 摄

大兴安岭区
小兴安岭区
呼伦贝尔区
松嫩平原区
三江平原区
阿尔泰山区
天山-准噶尔盆地西南部区
锡林郭勒草原区
长白山区
塔里木河流域区
库姆塔格区
西鄂尔多斯-贺兰山-阴山区
北京
黄渤海区
祁连山区
太行山区
羌塘-三江源区
秦岭区
六盘山-子午岭区
大别山区
黄山-怀玉山区
大巴山区
喜马拉雅东南部区
岷山-横断山北段区
鄱阳湖区
洞庭湖区
武夷山区
武陵山区
横断山南段区
南岭区
桂西黔南石灰岩区
西双版纳区
桂西南山地区
东海及台湾海峡区
南海区
南海区
海南岛中南部区
南海区
南海诸岛

Lower Hilly Region of South China

华南低山丘陵区

华南低山丘陵区位于中国南部，主要包括云南、海南、广西 3 个省（自治区）的 3 个生物多样性保护优先区域，分别为海南岛中南部区、西双版纳区和桂西南山地区，总面积 12 089.97 km^2。主要保护对象为热带雨林生态系统以及亚洲象、绿孔雀、望天树等珍稀濒危野生动植物。保护优先区内建有国家级自然保护区 18 处，总面积 7 251.24 km^2。

Lower Hilly Region of South China mainly consists of 3 priority areas of biodiversity conservation, namely the Central and South Hainan Island Area, Xishuangbanna Area and the Mountain Area of Southwest Guangxi, with a total area of 12,089.97 km^2, in 3 provinces (autonomous regions) including Yunnan, Hainan, Guangxi. The main protected objects are tropical rainforest ecosystems and rare and endangered species, such as *Elephas maximus*, *Pavo muticus*, and *Parashorea chinensis*. There are 18 national nature reserves in the priority areas, with a total area of 7,251.24 km^2.

热带雨林生态系统

Tropical Rainforest Ecosystem

热带雨林生态系统是华南低山丘陵区具有代表性的生态系统类型之一，现已建立尖峰岭、五指山、吊罗山、西双版纳等多个以该生态系统为主要保护对象的国家级自然保护区。

Tropical rainforest ecosystem is one of the representative ecosystem types in the Lower Hilly Region of South China. Many national nature reserves have been established, such as Jianfengling, Wuzhishan, Diaoluoshan, and Xishuangbanna, with this ecosystem as the main protection object.

奚志农 野性中国 / 摄

白头叶猴 国I CR

White-headed Black Langur
Trachypithecus leucocephalus

白头叶猴是第一个由中国人发现并命名的灵长类动物，是中国特有种，广西壮族自治区崇左市是其全球唯一的分布地。在其主要分布地建立了广西崇左白头叶猴国家级自然保护区。政府通过向农民租用耕地、种植食源性植物等措施，保障了白头叶猴的食物来源。此外，当地政府和社区还致力于恢复山脚植被、构建生态廊道，减轻栖息地碎片化对白头叶猴的影响。截至 2020 年，保护区内白头叶猴数量已从 20 世纪 80 年代的 300 余只上升至超过 1 200 只。

White-headed black langur, the first primate discovered and named by Chinese, is endemic to China, and found only in Chongzuo, Guangxi. Through such measures as renting farmland from farmers and planting food plants, the government managed to maintain food supply for the langurs. In addition, the local government and communities are also committed to restoring vegetations at the foot of mountains, building ecological corridors and reducing the impact of habitat fragmentation on white-headed langurs. By 2020, the number of white-headed langurs in the nature reserve has increased from 300-odd in 1980s to over 1,200.

程斌 / 摄

蒙有蔚 自然影像中国 / 摄

程斌 / 摄

程斌 / 摄

穿山甲 国I CR

Chinese Pangolin
Manis pentadactyla

穿山甲分布于中国长江以南地区。近年来，为保护这一物种，中国政府采取了一系列改革性措施：2018 年，全面停止商业性进口穿山甲及其制品；2020 年，《中国药典》将穿山甲除名；同年，穿山甲从国家二级重点保护野生动物提升为国家一级重点保护野生动物。监测野外种群、划定保护区、建立穿山甲保护研究中心和基因库等措施也正在逐步实施。近期，云南、广东、广西、海南等多地均有监测或救助野生穿山甲的记录。

Chinese pangolin is found in regions south of the Yangtze River. In recent years, to protect this species, the Chinese government has adopted a series of measures: in 2018, commercial import of pangolins and their products was completely banned; in 2020, Chinese pangolin was removed from *Chinese Pharmacopoeia*; in the same year, its national conservation level was upgraded from Category II to Category I. Meanwhile, other conservation measures were progressively implemented, such as monitoring wild populations, delineating nature reserves, and establishing pangolin conservation research centers and gene banks. Recently, wild Chinese pangolins have been monitored or rescued in multiple regions, including Yunnan, Guangdong, Guangxi, and Hainan.

李成 / 摄

董磊 西南山地 / 摄

董磊 西南山地 / 摄

亚洲象 国I EN

Asian Elephant
Elephas maximus

亚洲象是亚洲现存最大的陆生动物，在中国仅分布于云南省。随着云南西双版纳国家级自然保护区的建立以及亚洲象保护工作的实施，亚洲象种群数量已由 20 世纪 80 年代的 100 余头发展到如今的约 300 头。自 2009 年起，云南省地方政府为农民购买了野生动物肇事赔偿责任保险，并通过建立食物源基地、修建防象工程、探索使用无人机监测和发布预警信息等措施，在缓解人象冲突方面取得了一定成果。

Asian elephant is the largest terrestrial animal in Asia today. In China, it is found only in Yunnan Province. With the establishment of Xishuangbanna National Nature Reserve in Yunnan Province and the implementation of conservation efforts, the population of Asian elephants has grown from 100-odd in the 1980s to about 300 today. In order to alleviate conflicts between local people and elephants, in 2009, the local government of Yunnan Province introduced and covered compensation liability insurance for farmers with regards to economic loss caused by wildlife. At the same time, measures such as establishing food source bases, building facilities to ward off elephants, exploring the use of UAVs to monitor elephant activities and issue early warning information, have been proved successful.

董磊 西南山地 / 摄

董磊 西南山地 / 摄

董磊 西南山地 / 摄

绿孔雀 国I EN

Green Peafowl
Pavo muticus

绿孔雀在中国仅分布于云南省。2009 年，云南省公布了《云南省极小种群物种拯救保护规划纲要和紧急行动计划（2010—2020）》，绿孔雀被列为优先保护物种。对其实施种群监测体系建设、栖息地管控、补水点和食源地建设等恢复与改造项目。民间机构积极参与到绿孔雀的科普宣传与监测保护工作中，通过聘任绿孔雀分布地周边的农民作为巡护员的方式，推进社区保护工作的实施。

In China, green peafowl is found only in Yunnan Province. In 2009, Yunnan published Planning Outline and *Emergency Action Plan of Yunnan Province on Rescuing and Conserving Species with Extremely Small Populations (2010-2020)*, in which green peafowl was listed among the protected species with conservation priority. Restoration projects were carried out, including the development of a population monitoring system, habitat control, and construction of water supply spots and food sources. Wild China Film, Friends of Nature and other non-governmental organizations were also involved in the monitoring and conservation of green peafowl, as well as relevant public education efforts. Community protection work is promoted through such measures as hiring local farmers to patrol the habitats of green peafowl.

奚志农 野性中国 / 摄

奚志农 野性中国 / 摄

陈建伟 自然影像中国 / 摄

望天树 国I EN

Parashorea chinensis

望天树是东南亚热带雨林的标志物种，在中国仅分布在广西壮族自治区和云南省。1974 年，包括望天树在内的龙脑香科植物类群首次在云南勐腊被发现，望天树也得到了正式命名。在望天树的分布地已建立保护区进行就地保护，中科院昆明植物研究所热带植物园亦有栽培。

Parashorea chinensis is a symbolic Southeast Asian tropical rainforest plant. In China, it is found only in Guangxi and Yunnan Province. Camphoraceae plants, including *Parashorea chinensis*, were first discovered in 1974 in Mengla, Yunnan Province, and *Parashorea chinensis* in this area was then officially named. Nature reserves have been established within the distribution range of this species, which is also cultivated in Tropical Botanical Garden of the Kunming Institute of Botany, Chinese Academy of Sciences.

陈建伟 自然影像中国 / 摄

刘冰 / 摄

大兴安岭区
呼伦贝尔区
小兴安岭区
松嫩平原区
三江平原区
阿尔泰山区
天山-准噶尔盆地西南部区
锡林郭勒草原区
长白山区
塔里木河流域区
库姆塔格区
西鄂尔多斯-贺兰山-阴山区
北京
黄渤海区
祁连山区
太行山区
羌塘-三江源区
秦岭区
六盘山-子午岭区
大别山区
黄山-怀玉山区
大巴山区
岷山-横断山北段区
喜马拉雅东南部区
鄱阳湖区
洞庭湖区
武夷山区
东海及台湾海峡区
横断山南段区
武陵山区
南岭区
桂西黔南石灰岩区
西双版纳区
桂西南山地区
南海区
南海区
海南岛中南部区
南海诸岛

Marine and Coast

海洋与海岸

海洋与海岸生物多样性保护优先区域主要包括 3 个区域，分别为黄渤海保护区域、东海及台湾海峡保护区域和南海保护区域。主要保护对象为珊瑚礁生态系统以及西太平洋斑海豹、绿海龟、中华白海豚等珍稀濒危野生动物。中国已建立各类国家级海洋自然保护区 32 处。

Priority areas of marine and coast biodiversity conservation includes three main areas, namely the Yellow Sea and Bohai Sea Protected Region, the East China Sea and Taiwan Strait Protected Region, and South China Sea Protected Region, of which the conservation subjects are coral reef ecosystems and rare or endangered species, such as *Phoca largha*, *Chelonia mydas*, and *Sousa chinensis*. China has established 32 national marine nature reserves of different types.

珊瑚礁生态系统

Coral Reef Ecosystem

珊瑚礁生态系统是热带海洋最突出、最具有代表性的生态系统类型，被认为是具有重要服务功能的全球生态系统之一。中国已建立徐闻珊瑚礁、三亚珊瑚礁、铜鼓岭等多个以该生态系统为主要保护对象的国家级自然保护区。

Coral reef ecosystems are the most prominent and representative types of ecosystem in tropical seas, considered as a global ecosystem with important functions. China has established national nature reserves focusing on the conservation of coral reef ecosystems, such as Xuwen Coral Reef, Sanya Coral Reef, and Tongguling National Nature Reserve.

苏羽 / 摄

西太平洋斑海豹 国Ⅰ LC

Spotted Seal
Phoca largha

西太平洋斑海豹是唯一能在中国海域繁殖的鳍足类动物，在中国繁殖于渤海辽东湾。1992 年，大连建立了首个以西太平洋斑海豹及其栖息地为主要保护对象的保护区，并于1997 年晋升为国家级自然保护区。2002 年，辽宁大连斑海豹栖息地湿地被评为国际重要湿地。政府及民间机构积极开展西太平洋斑海豹的种群监测、保护、救助放归及人工繁育工作。近年来，已有百余只西太平洋斑海豹被救助并放归，同时对部分放归个体开展了卫星跟踪监测。2021 年，新调整的《国家重点保护野生动物名录》公布，西太平洋斑海豹升级为国家一级重点保护野生动物。

Spotted seal, the only fin-footed animal that breeds in the maritime territory of China, reproduces in Liaodong Bay of the Bohai Sea. The first nature reserve for spotted seals and their habitats was established in Dalian in 1992, and was upgraded to a national nature reserve in 1997. In 2002, the Dalian Spotted Seal Habitat Wetland of Liaoning Province was entitled as Wetland of International Importance. The government and non-governmental organizations are also actively involved in population monitoring, protection, rescue and release and captive breeding efforts of this species. In recent years, more than 100 spotted seals have been rescued and released, some of which under satellite tracking and monitoring.
In 2021, the newly adjusted “List of National Key Protected Wildlife" was published, and the species was listed as a Class I.

王小平 / 摄

张程皓 / 摄

刘思阳 西南山地 / 摄

顾晓军 自然影像中国 / 摄

中华白海豚 国I VU

Indo-Pacific humpback dolphin
Sousa chinensis

中华白海豚在中国主要分布于东南沿海诸省的近海海域，是中国海域遇见率最高的鲸豚。中国政府高度关注这一物种，在中国现存最大的中华白海豚栖息地建立了珠江口中华白海豚国家级自然保护区，港珠澳大桥的设计与建设也高度重视对该物种的保护。2019 年，在国庆 70 周年大阅兵中，中华白海豚的身影出现在了“绿水青山”方阵中。

Indo-Pacific humpback dolphin, mainly found in the offshore waters of several southeast coastal provinces in China, is the most commonly encountered species of the family Lissodelphis in the maritime territory of China. The national government pays close attention to this species and has established Pearl River Estuary National Nature Reserve for Indo-Pacific Humpback Dolphins in the largest existing habitat of the species in the country. Conservation of this species is also a high priority in the design and construction of the Hong Kong-Zhuhai-Macao Bridge. In 2019, during the military parade on China's 70th National Day, an image of Indo-Pacific humpback dolphin appeared in the Lucid Waters and Lush Mountains formation.

冯抗抗 / 摄

吴哲浩 / 摄

陈建伟 自然影像中国 / 摄

中华凤头燕鸥 国I CR

Chinese Crested Tern
Thalasseus bernsteini

中华凤头燕鸥也被称为“神话之鸟”，仅在东海和黄海繁殖。在消失了 60 多年后，2000 年，该物种在中国福建沿海的马祖列岛再次被发现；2004 年，在浙江韭山列岛又发现了一个繁殖种群。2013 年，浙江自然博物馆研究团队开始与国外科学家合作，在浙江韭山列岛国家级自然保护区内开展该物种的人工招引试验，并取得初步成功。从 2015 年开始，招引工作在象山韭山列岛和定海五峙山列岛同时开展，并连年取得成功。目前，该物种已从 2010 年的不足 50 只增长到超过 100 只。2021 年，新调整的《国家重点保护野生动物名录》公布，中华凤头燕鸥升级为国家一级重点保护野生动物。

Chinese crested tern, also known as “Bird of Myth”, breeds only in the East China Sea and the Yellow Sea. After a disappearance of more than 60 years, this species was discovered once again on Mazu Islands off the coast of Fujian in 2000. In 2004, another breeding population was found on Jiushan Islands in Zhejiang. Since 2013, a research team from Zhejiang Natural History Museum has been working with foreign scientists in experiments to attract this species in Jiushan Islands National Nature Reserve, which were preliminary successful. Since 2015, efforts to attract the species have been simultaneously carried out on Jiushan Islands in Xiangshan and Wuzhishan Islands in Dinghai. The efforts have been continually successful. So far, the population of this species has increased from less than 50 in 2010 to more than 100. In 2021, the newly adjusted “List of National Key Protected Wildlife” was published, and the species was listed as a Class I.

冯江 自然影像中国 / 摄

陈水华 / 摄

冯江 自然影像中国 / 摄

冯江 自然影像中国 / 摄

绿海龟 国 I EN

Green Turtle
Chelonia mydas

绿海龟在中国分布于江苏、浙江、福建、台湾、广东等地的沿海地带，产卵场仅在福建西部和广东东部的沿岸和岛屿。2010 年，广东惠东港口海龟国家级自然保护区启动绿海龟人工繁殖项目，并于 2017 年攻克了海龟人工繁殖技术难关，填补了中国在绿海龟全人工繁殖技术上的空白。近年来，中国积极开展绿海龟放流工作，2019 年 1 月放流 100 只，其中包括救助、罚没以及人工繁殖个体，2020 年放流 141 只。2021 年，新调整的《国家重点保护野生动物名录》公布，绿海龟升级为国家一级重点保护野生动物。

Green turtle is found in coastal areas of China, including Jiangsu, Zhejiang, Fujian, Taiwan, and Guangdong, while it spawns only on the coasts and islands of western Fujian and eastern Guangdong. Huidong National Reserve of Sea Turtle in Guangdong Province launched a green turtle captive breeding project in 2010 and overcame technological difficulties in 2017, filling the gap in the full-artificial breeding technology of green turtles in China. In recent years, China has been actively restoring the wild population of green turtle. In January 2019, 100 turtles were released, including the rescued, confiscated and captive-bred individuals. 141 turtles were released into the sea in 2020. In 2021, the newly adjusted “List of National Key Protected Wildlife” was published, and the species was listed as a Class I.

陈建伟 自然影像中国 / 摄

苏羽 / 摄

苏羽 / 摄

苏羽 / 摄

蛇岛蝮 国II VU

Shedao Island Pit Viper
Gloydius shedaoensis

蛇岛蝮是中国特有种，于 1979 年被中国科学家正式命名，主要分布在辽宁省的蛇岛上。1980 年，以蛇岛蝮等野生动物及其栖息地为主要保护对象的辽宁蛇岛老铁山国家级自然保护区被批准建立，这也是辽宁省的第一个国家级自然保护区。保护区科研人员积极开展蛇岛蝮的种群监测以及科学研究工作，登岛打柴、捕蛇等非科学研究活动则被全面禁止。2021 年，新调整的《国家重点保护野生动物名录》公布，该物种被列为国家二级重点保护野生动物。

Shedao island pit viper, an endemic viper species in China officially named by Chinese scientists in 1979, is mainly found on Snake Island in Liaoning, China. In 1980, the Laotieshan National Nature Reserve of Snake Island was approved to be established with an aim of protecting wild animals, including Shedao island pit viper, and their habitats. This is also the first national nature reserve in Liaoning Province, where all non-scientific research activities such as logging and snake-catching are prohibited. Scientific researchers in the reserve take an active part in population monitoring and scientific research on Shedao island pit viper. In 2021, the newly adjusted "List of National Key Protected Wildlife" was published, and the species was listed as a Class II.

王小平 / 摄

陈建伟 自然影像中国 / 摄

伟 自然影像中国 / 摄

后记

1972 年，联合国第一次人类环境会议召开。会议通过了《人类环境宣言》，号召各国政府和人民为保护和改善环境而奋斗，它开创了人类社会环境保护事业的新纪元，这是人类环境保护史上的第一座里程碑。20 年后，1992 年，联合国环境与发展大会通过《21 世纪议程》，《生物多样性公约》开放签署。如今，中国已成为全球生态文明建设的重要参与者、贡献者、引领者。

本图册的编著，使我们深切感受到近 30 年来，中国为履行生态环境相关多边公约和其议定书所做出的艰苦努力，也见证了生物多样性保护工作和生态文明建设取得的历史性成就。在此，愿对一直关心和支持中国生物多样性保护工作的国际组织和机构、中央和地方各级政府部门、企业、行业协会和学会、高校、科研院所、社会组织以及社会公众等利益相关方表示衷心的谢意！

人类是一个整体，地球是一个家园。“生态文明：共建地球生命共同体”既是《生物多样性公约》第十五次缔约方大会（COP15）的主题，也是人类对未来的美好寄语。愿我们坚持尊重自然、顺应自然、保护自然，同心协力为推动全球生物多样性治理进程，增进人类福祉作出新贡献。

Afterword

The first United Nations Conference on the Human Environment was held in 1972. The conference adopted the "Declaration of the United Nations Conference on the Human Environment", calling on governments and people of all countries to strive for the protection and improvement of the environment. It inaugurated in a new era of environmental protection in human society, which is the first milestone in the history of human environmental protection. Twenty years later, in 1992, the United Nations Conference on Environment and Development adopted "Agenda 21" and the "Convention on Biological Diversity" was opened for signature. Today, China has become an important participant, contributor and leader in the process of global ecological civilization construction.

In the process of compiling this album, we have been deeply aware of the arduous efforts China has made in the past three decades to implement the multilateral conventions and its protocols related to the ecological environment, and also, we have witnessed the historic achievements of biodiversity protection and ecological civilization construction. Here, we would like to express our heartfelt gratitude to the international organizations and institutions, central and local government departments, enterprises, industry associations and societies, universities, scientific research institutes, social organizations, the public and other stakeholders who have been always caring about and supporting biodiversity conservation in China!

The mankind is a whole, and the earth is a home. "Ecological Civilization: Building a Shared Future for All Life on Earth" is not only the theme of the COP 15, but also a good wish for the future of mankind. We hope that we will continue to respect, conform and protect nature, and work together to make new contributions to promoting the global biodiversity management and enhancing human well-being.

图书在版编目（CIP）数据

中国的生物多样性保护：汉英对照 / 中华人民共和国生态环境部编著. -- 北京：中国环境出版集团，2021.12
ISBN 978-7-5111-4855-1

Ⅰ. ①中… Ⅱ. ①中… Ⅲ. ①生物多样性 - 生物资源保护 - 中国 - 图集 Ⅳ. ①X176-64

中国版本图书馆CIP数据核字(2021)第184574号

审图号：GS(2021)6103号

支撑团队　中国国家地理自然教育中心
支　　持　全球环境基金中国国家公园体制机制创新项目

gef　UNDP　FECO　C-PAR

出 版 人　武德凯
策划编辑　王素娟
责任编辑　曹　玮
封面设计　董　良
封面照片　冯　江

出版发行　中国环境出版集团
（100062 北京市东城区广渠门内大街 16 号）
网　　址：http://www.cesp.com.cn
电子邮箱：bjgl@cesp.com.cn
联系电话：010-67112765（编辑管理部）
发行热线：010-67125803，010-67113405（传真）
印　　刷　北京雅昌艺术印刷有限公司
经　　销　各地新华书店
版　　次　2021 年 12 月第 1 版
印　　次　2023 年 12 月第 2 次印刷
开　　本　787×1092 1/16
印　　张　12.75
字　　数　100 千字
定　　价　368.00 元